Lecture Notes in Earth Sciences

Lecture Notes in
Earth Sciences

Edited by Somdev Bhattacharji, Gerald M. Friedman,
Horst J. Neugebauer and Adolf Seilacher

24

Bernhard Knipping

Basalt Intrusions
in Evaporites

Springer-Verlag
Berlin Heidelberg GmbH

Author

Dr. Bernhard Josef Knipping
Institut für Mineralogie und Mineralische Rohstoffe
Fachgebiet Salzlagerstätten und Untergrund-Deponien
Technische Universität Clausthal
Adolph-Roemer-Str. 2 A, D-3392 Clausthal-Zellerfeld, FRG

Translated by
Ralph B. Phillips
Zeiss-Str. 17a, D-8520 Erlangen, FRG

ISBN 978-3-540-51308-7   ISBN 978-3-540-46193-7 (eBook)
DOI 10.1007/978-3-540-46193-7

© Springer-Verlag Berlin Heidelberg 1989
Originally published by Springer-Verlag Berlin Heidelberg New York in 1989

2132/3140-543210 – Printed on acid-free paper

Preface

Various salt deposits found worldwide have experienced a phase of volcanic activity during their history. In addition to the economic significance of the alteration and redistribution processes occurring during volcanism such deposits aid in understanding the interaction between basalts and salt rocks.

During the Miocene (about 10 - 25 Ma) basaltic melts intruded into the Zechstein evaporites of the Werra-Fulda region (Germany), which formed more than 250 Ma ago. They reached the surface at only a few points and usually solidified at greater depths. Today, about 15 NS-striking systems of basaltic dikes are exposed at the Hattorf potash mine. Mobile components such as H_2O and CO_2 found their way into the neighbouring salt rocks during and following intrusion of the magmas. Thus, extensive mineral reactions have a close genetic relationship to the basalt volcanism. These mineral reactions led to the alteration of entire deposits, above all the K-Mg rocks of the potash seams, which are particularly sensitive to hydrous solutions and increased temperature. During these processes gaseous mixtures were fixed in partially great amounts in chloride rocks through recrystallization of salt minerals. Native sulfur was also found occasionally at and near basalt dikes.

Hence the sulfate type potash salts of the Werra-Fulda mining district are particularly suitable for studies concerning evaporite and basalt genesis as well as problems concerning underground disposal of heat-producing toxic wastes (natural analogue).

I would like to sincerely thank Prof. Dr. A.G. Herrmann for initiating this study and for his continuous support through discussions. He also enabled the K-Ar age dating of five basalt samples.

The scientific work was made possible by a grant from the German Research Council (DFG), to whom I am particularly grateful.

For the extensive sampling in the Hattorf mine I had the friendly support of the main administration of Kali und Salz AG and the management at the Hattorf mine. Kali und Salz AG also enabled three K-Ar whole-rock datings. I would like to thank above all Dr. K.-Ch. Käding, H. Klee, and Dr. G. v.Struensee for their support during sampling and many discussions.

I am grateful to Dr. G. Delisle (Federal Institute for Geosciences and Natural Resources, Hanover) for an introduction into the mathematical fundamentals of temperature calculations and further discussions. I am very thankful to H. Peters and Dr. G. Siebrasse (Institute for Numerical and Applied Mathematics, University of Göttingen) for their help in modelling a computer program for my specific purposes. I thank H. Laschtowitz (Institute of Geochemistry, University of Göttingen) for his patience in correcting the numerous 'breakdowns' I caused on the institute computer. During the evaluation of the data I received valuable information from Prof. Dr. M. Rosenhauer (Institute of Mineralogy and Petrology, University of Göttingen).

The Society for Scientific Data Processing (GWD, Göttingen) helped with the calculation and working time.

The entire facilities at the Institute of Geochemistry (University of Göttingen) were at my disposal during the experimental studies and their evaluation. I would particularly like to thank Ms. I. Hinz, Ms. B. Dietrich, Ms. G. Mengel, Mr. H. Herborg, Mr. K. Herrmann (Institute of Mineralogy and Mineral Resources, Technical University of Clausthal-Zellerfeld), Prof. Dr. J. Hoefs, Dr. K. Mengel, Dr. H. Nielsen, Prof. Dr. K.-H. Nitsch, Mr. E. Schiffczyk, Mr. R. Przybilla, and Prof. Dr. K.H. Wedepohl. Dr. H. Nielsen also kindly provided some unpublished sulfur isotope data for comparative purposes.

The radiation and preparation of the samples for neutron-activation analysis were helpfully carried out in the Triga I research reactor of the Medical University of Hanover and the Isotope Laboratory for Biological and Medical Research of the University of Göttingen.

I would like to thank Mr. R.B. Philipps for translating this study. I must also express my thanks to the Springer-Verlag, and especially to Dr. W. Engel for his support.

My friends and colleagues provided me with a comfortable working environment and constantly renewed motivation. I am expecially thankful to Dr. Andreas Kirchhoff, Dr. Heide Peters, Dr. Bernhard Schnetger, Wolfgang Schultz, and Lieselotte E. von Borstel.

This work is dedicated to my family, Claudia, Katrin, Friederike, and Maren, who had to do without me, especially in the last months of work.

Contents

1 Introduction

The evaporite deposits of the Werra district, especially in the Hattorf mining field, are considered a worldwide unique location for the occurence of numerous basalt dikes and magmatic fluid phases fixed in salt rocks. In spite of the great number of studies dealing with the magmatites in the Werra region, previous investigations have rarely attempted more than a predominantly 'qualitative' description of the basaltic rocks and the effects of volcanism on the evaporites (see Chapter 2). The method of interpreting the mineralogical and chemical composition of the evaporites at the basalt contact is based on previous works (KNIPPING 1984; KNIPPING & HERRMANN 1985). This study should contribute to understanding (i) the mechanism of intrusion of the basaltic melts and (ii) the metamorphic processes occurring in the evaporites caused by mobile phases during volcanism. Hence, the following methods were applied:

- The mineralogical and chemical description of the basaltic rocks with recent nomenclature including the possible differences between individual dikes and between surface- and subsurface-exposed basalts. Seven surface and 48 subsurface exposures at the Hattorf mine of Kali & Salz AG were studied.
- Application of the most recent knowledge on basalt genesis for interpreting observational and experimental results.
- Studies on the sulfur and carbon isotope distributions of the native sulfur from several subsurface exposures and the enrichments of gases (predominantly CO_2) in the evaporites.
- Calculation of the spatial and temporal temperature distribution in the evaporite rocks following intrusion of the basaltic melts.

For purposes of clarity a few of the terms which will be used frequently here will first be defined:

basalt - all of the intrusive rocks studied can be assigned mineralogically and chemically to the basalt family in a broader sense. Thus, the terms *basaltic rock* or, in short, *basalt* will be used for these rocks.

rock salt - instead of the term *salt* for halitic rocks the term *rock salt* is used. Besides, the evaporites are generally designated as *host rocks* (for the basalt dikes) as well.

gases - especially in the German literature the term *carbon dioxide* or *carbonic acid* (= *Kohlensäure*) is frequently used for the gases enclosed in the evaporites of the Werra-Fulda district. ACKERMANN et al (1964) found, in addition to carbon dioxide, considerable amounts of nitrogen and minor amounts of methane. In the following therefore the terms *gas mixture* or *gas* will be used.

The various basalt dikes found in the Hattorf mining field are described here in terms of their mineralogy and geochemistry for the first time. In doing so it is necessary to number them from east to west. To avoid confusion with older numerations (e.g. SIEMENS 1971) the various dike systems are designated by capital letters (<A> to <P>).

2 Geological setting and petrography of evaporites and basaltic rocks in the Werra mining district, Germany

2.1 Geological setting and petrography

About 250 Ma ago (Upper Permian; MENNING 1986) extensive portions of present central and western Europe were covered by the Zechstein sea. The first evaporites were deposited in the Marginal basins of this sea. Since the Werra-Fulda region (Germany) in the eastern part of the Hessian basin belonged to these marginal basins, the saline cycle Zechstein 1 (Z1) is also named the *Werra series*. The Werra-Fulda potash district of today is bordered to the north and east by the Richelsdorf Mountains and Thüringen Forest, respectively. To the south it borders on the basalt masses of the Rhön and Vogelsberg (fig. 2.1). The 1000 km² Werra district is separated from the 100 km² Fulda district by the Fulda-Großenlüder graben. The flat lying evaporite beds of the Werra series dip to the SW with an average of 2° - 3°.

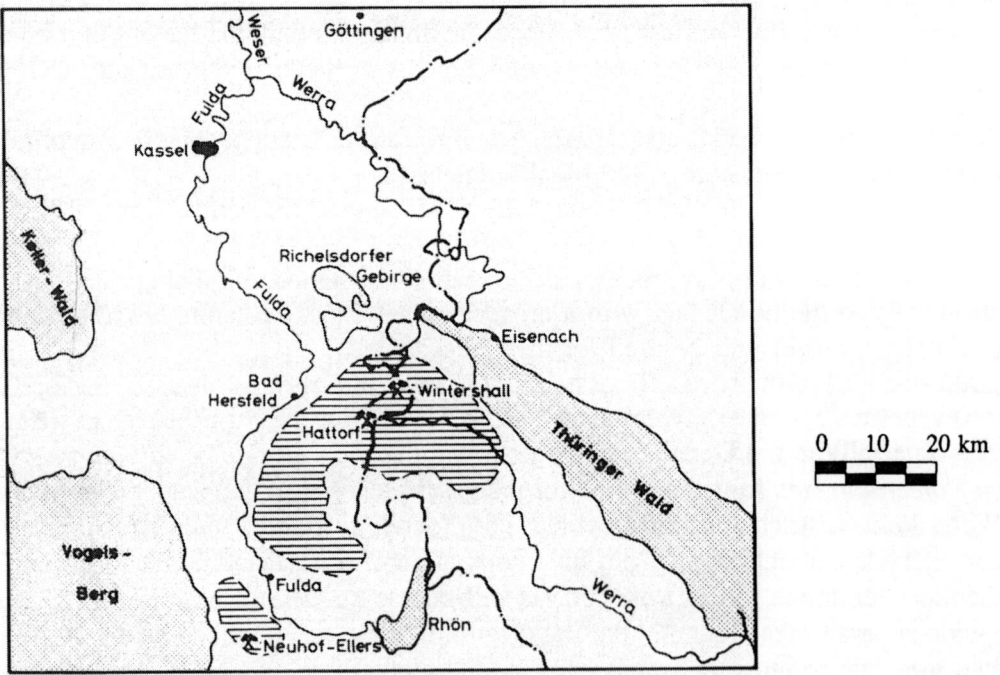

Fig. 2.1 The Werra-Fulda district with the Wintershall, Hattorf and Neuhof-Ellers potash mines (after MESSER 1978). The distribution of the potassium-bearing salt rocks is hatched.

K-Mg minerals formed repeatedly within the first saline cycle. *Thüringen* (K1Th) and *Hessen* (K1H) at depths of about 800 m are the main potash seams presently being mined in the Werra ore district. These seams form the upper and lower boundaries of the 100 to 150 m-thick beds of the Werra rock salt (Na1). Stratigraphically above K1H there are six accompanying seams (*Begleitflöze*), which are partially of economic value due to their thicknesses. In the center of the Hattorf mining field there is still another potash salt seam about 20 m above K1H. The genesis of the so-called *Hattorf* seam has not yet been explained. The stratigraphy of the Zechstein 1 in the Werra-Fulda basin is given in table 2.1.

Tab. 2.1 Stratigraphy of the Zechstein in the Werra-Fulda basin (KÄDING 1978).

Fulda potash region			Werra potash region		
m			m		
			0 - 2	Oberer Werra-Anhydrit	A1rβ
			0 - 15	Oberstes Werra-Steinsalz	Na1r
3 - 7	Oberer Werra-Anhydrit	A1r	5 - 7	Oberer Werra-Anhydrit	A1rα
8 - 12	Oberer Werra-Ton	T1r	8 - 10	Oberer Werra-Ton	T1r
70	Oberes Werra-Steinsalz	Na1γ	90 - 150	Oberes Werra-Steinsalz	Na1γ
2 - 3	Kaliflöz Hessen	K1H	2 - 3	Kaliflöz Hessen	K1H
35 - 55	Mittleres Werra-Steinsalz	Na1β	50 - 60	Mittleres Werra-Steinsalz	Na1β
2 - 3	Kaliflöz Thüringen	K1Th	2 - 10	Kaliflöz Thüringen	K1Th
60 - 100	Unteres Werra-Steinsalz	Na1α	80 - 110	Unteres Werra-Steinsalz	Na1α
3	Unterer Werra-Anhydrit	A1	3 - 25	Unterer Werra-Anhydrit	A1
5 - 8	Anhydritknotenschiefer	CaA1	6 - 9	Anhydritknotenschiefer	CaA1
6 - 20	Werra-Karbonat (Zechsteinkalk)	Ca1	6 - 15	Werra-Karbonat (Zechsteinkalk)	Ca1
0.2 - 0.5	Unterer Werra-Ton (Kupferschiefer)	T1	0.3 - 0.5	Unterer Werra-Ton (Kupferschiefer)	T1
0 - 4	Werra-Konglomerat (Zechsteinkonglomerat)	C1	0 - 3	Werra-Konglomerat (Zechsteinkonglomerat)	C1

The potash-bearing salt rocks of the Werra series are among the 0.5 % to 5 % of the salt deposits worldwide that belong to the *sulfate type* of marine evaporites (HERRMANN 1987, 1989). Rocks of the sulfate type form from evaporating seawater, which, however, has lost part of its $MgSO_4$ content. In Z1 and Z2 the brines obviously contained only one-fourth to one-half of their original $MgSO_4$ content theoretically expected in unaltered seawater. The potash salt seams differ in their average mineralogical composition, above all in the amounts of the main components halite, sylvite, carnallite and kieserite (e.g., WEBER 1961; ROTH & MESSER 1981; KOKORSCH & PSOTTA 1984). In the Thüringen seam the so-called *Trümmer carnallitite* overlies a kieseritic Hartsalz. The Hessen seam is composed predominantly of kieseritic Hartsalz and in some parts of carnallitite as well. Certain potassium-bearing salt rocks contain the highest kieserite contents of the sulfate-type salt deposits known in the world (GIMM 1968). For example, the *Flockensalz* in K1H can contain > 60 % kieserite (BRAITSCH 1962, p. 133; 1971, p.175). The amounts of $MgSO_4$ are of considerable economic value because the potash salt occurrences of the Werra-Fulda potash district belong to the few in the world which enable the production of potassium sulfate as fertilizer.

In the Tertiary (Miocene) the Hessian Depression was a site of intensive volcanism. Basaltic melts intruded into the evaporites of the Werra-Fulda district. South of the present Hattorf mining field basalt cupolas, some of great dimensions, are exposed at the surface. However, the number of surface exposures decreases sharply northward. In the ca. 110 km² field area of this study basalts occur relatively rarely at the surface. Yet, in the underground mines at depths of about 800 - 900 m they are exposed as dikes and sills, some of which are several kilometers long. The basalts are mostly bound to NNE-SSW- and NNW-SSE-striking fracture systems. Figure 2.2 shows the surface and subsurface basalt occurrences in the area of the Hattorf mine and the outcrops sampled for this study.

The largest studied occurrence at the surface is Soisberg, which is about 0.28 km². All of the basalts exposed at the surface collectively cover an area of around 0.4 km². This is less than 0.4 % of the area shown on the map (fig. 2.2). The area taken up by the subsurface basalts was calculated to be 0.005 - 0.01 km², assuming that the dikes have average thicknesses of 0.5 m and lengths of about 10 - 20 km. This corresponds to about 1.25 % to 2.5 % of the area covered by basalts at the surface. The area ratio between the surface and subsurface basalts is consequently about 80:1 to 40:1. The order of magnitude of this rough estimation is surely realistic although exact data on the lengths of the exposed subsurface basalt dikes were not available.

In underground exposures the basalt dikes are frequently interrupted: they end at one point and begin again a few meters west or east of this point. Positions of the dikes are frequently lacking completely (fig. 2.3). The contact between the basalt and rock salt is usually sharp, showing no clear effects on the host rock. On the level of K1Th and K1H and the accompanying seams, on the contrary, the basaltic melts have frequently penetrated sideways into the potash horizons over distances of tens of meters up to > 100 m.

Pinnate cracks and joints oriented parallel to the dikes are observed in the underground exposures. They are evidence of brittle deformation, which has also led to the formation of joints in the rock salt and potash horizons. In most cases, the cracks have been healed by secondary minerals, not basalt.

Rock alterations, in part extensive, and accumulations of gas in the evaporites (*Knistersalz*, cf. chapter 6) are obviously related to the basalt intrusions. Mixtures of gas occur primarily in the potash horizons. According to TAMMANN & SEIDEL (1932) and KÜHN (1951) they are under a pressure of 0.78 - 1.2 MPa. These mineral-bound gases are released by blasting during tunnel driving. Round voids, which frequently extend through the rock as tubes, form at the points of exit. Here, the salt rock

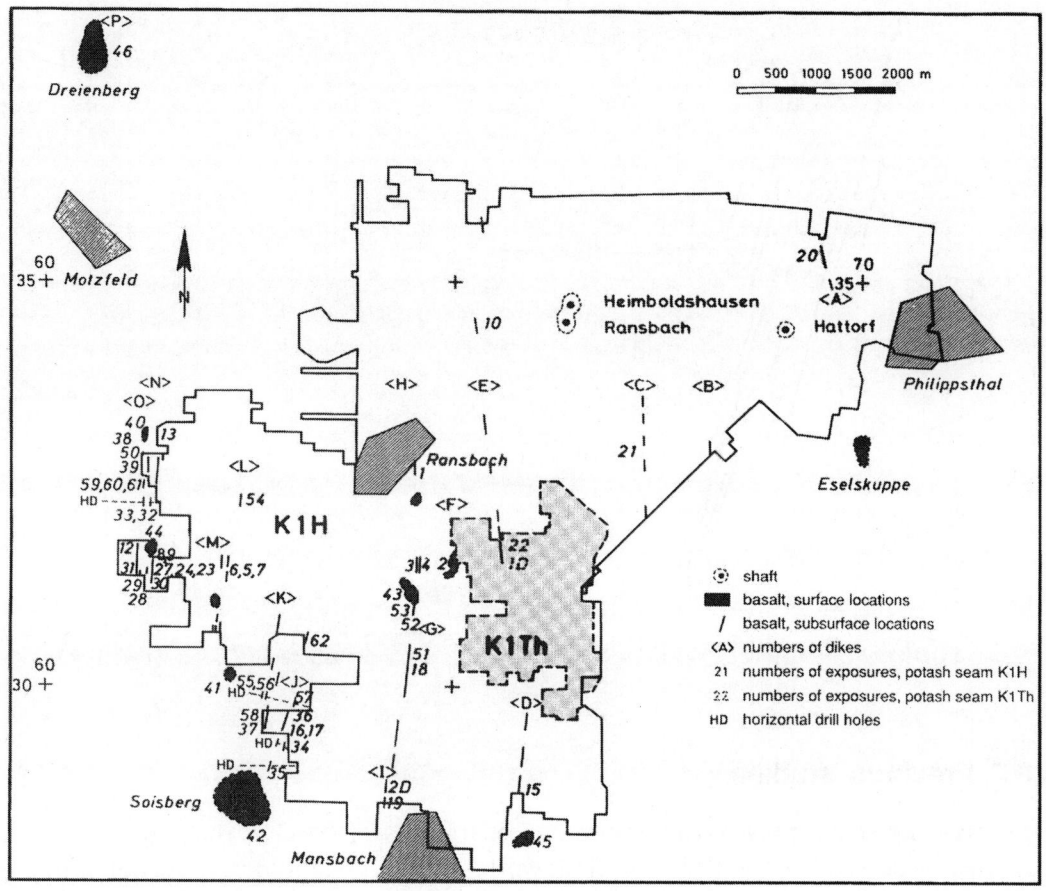

Fig. 2.2 Surface and subsurface basalt occurrences in the area of the Hattorf potash mine and the outcrops sampled for this study.

frequently displays a typical fissility. At the Hattorf mine a maximum of 25 000 t of salt rock were ejected and 200 000 to 250 000 m³ of gas were released during separate gas explosions (KÄDING 1975).

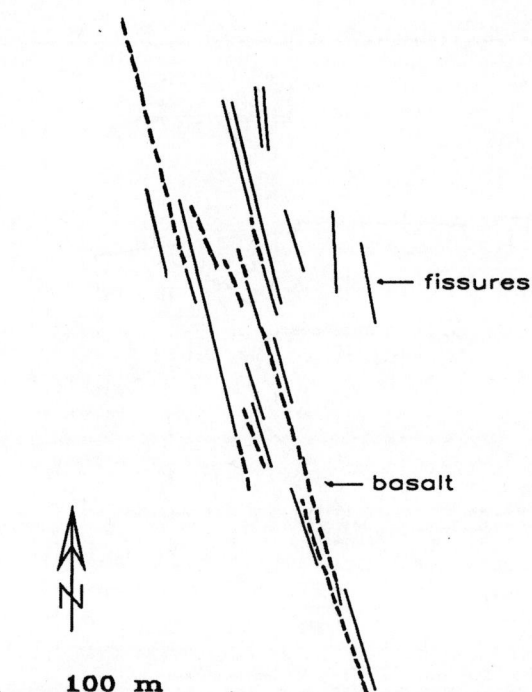

Fig. 2.3 Basalt dike on the level of the Hessen potash seam (JAHNE et al 1983).

2.2 Previous studies

In this chapter the previous literature on occurrences of basaltic rocks in evaporites will be discussed briefly. Some of these works will be gone into in more detail in the appropriate chapters.

Basaltic rocks occurring at the surface in the area of the Hattorf mine were first identified in 1877 by v. KOENEN (1886, 1888). BÜCKING (1881) described them according to the nomenclature of that time as tephrite, basanite, feldspathic basalt, basanitoid, and nepheline basalt. CO_2-bearing gas mixtures had already been encountered in the Werra region during the initial exploratory drillings for potash salts (FRANTZEN 1894). Shortly after mining at Hämbach had begun (1901, Kaiseroda I shaft), basalt dikes and releases of gas were observed. (NAUMANN 1911, 1914; SCHEERER 1911; BECK 1912a, b; GROPP 1919). GRUPE (1913) first recognized that basaltic rocks can occur as dikes and sills.

DIETZ (1928b, p. 43) presented the first data on the intrusion mechanism of basaltic melts. Due to the folding observed in the evaporites near a dike in the Sachsen-Weimar mine DIETZ concluded that the magmas must exert considerable force during intrusion. He also pointed out that a fluid phase must have participated in the fixation of the gas mixtures in the evaporites (DIETZ 1928a, b). BESSERT (1933) reported on enrichments of anhydrite, kainite, and polyhalite at the basalt contact.

The accumulations of gas in the evaporites became a technical and economical problem. Therefore, several studies dealing with the fixation of CO_2-bearing gas mixtures in the salt rocks, which is genetically related to the basalts, followed (e.g., HARTWIG 1954; KÜHN 1951; BAAR 1958; MÜLLER 1958; HOPPE 1958, 1960; OELSNER 1961; D'ANS 1967). KÄDING (1962) did the first comprehensive geological study of the basalts occurring in the Werra region. SIEMENS (1971) published results of magnetic ΔT measurements taken during surface reconnaissance of the basalts in the Werra potash district. More recent petrographical studies and several chemical analyses of the surface basalts south of the Hattorf mine are given in LAEMMLEN & MEISL (1975).

The basaltic rocks in the eastern portion of the Werra region and their effects on the salt deposit were recently described by, for example, FRANKE (1974), KOCH (1978), and KOCH & VOGEL (1980). However, these studies are in part methodically incomplete. The conclusions drawn from the results are unfortunately based frequently on presumptions and do not take into account the current knowledge on the genesis of marine evaporites and basaltic rocks (cf. chapter 9). Using various reaction models for isothermal and polythermal conditions KNIPPING (1984) and KNIPPING & HERRMANN (1985) were the first to quantitatively determine the material transport and mineral reactions during alteration of a carnallitic rock at the contact to an olivine nephelinite (Hattorf mine). GUTSCHE (1987) and GUTSCHE & HERRMANN (1988) developed in expanded form similar model calculations for a basalt outcrop in K1H.

The geologically interesting occurrence of basalts in evaporites is also known from other salt deposits. WIMMENAUER (1952), BRAITSCH et al (1964) and HURRLE (1976) are important works dealing with the olivine nephelinites ('ankaratrites') in the salt deposits of Buggingen (southern Baden, West Germany).

The alkali-olivine basalts exposed in the Permian evaporites of the Salado Formation near Carlsbad, New Mexico, were investigated by LOEHR (1979) and BROOKINS (1981, 1984, 1986), for example. However, these works are also partially deficient in terms of methodology and mineralogy, as far as the genesis of evaporites is concerned.

The occurrence of volcanics in the hardly studied Cretaceous evaporites of the Khorat Plateau (Thailand) is shown in HITE & JAPAKASETR (1979, fig. 2).

3 Exposures and sampling

The geographical location of the exposures is given in fig. 2.2. Nearly all subsurface samples were taken from horizon 1 (at the K1H level) in the Hattorf mine of Kali & Salz AG. Mining here has progressed much further than in horizon 2 (at the K1Th level). Since a great number of the old workings and tunnels are no longer accessible, the selection of sites for sampling was restricted to the tunnels and sections of the mine which are presently used and exploited. Due to the poor exposures, dike <B> was not able to be sampled. Most of the outcrops described in this study have already become inaccessible in the meantime.

The photos in this chapter were taken at the stratigraphic level K1H. The photographic documentation of the outcrops was hindered by the extensive debris cover. Therefore, out of the great number of outcrops only a selection of the various forms of subsurface basalt occurrence can be shown. All outcrops studied are listed in table 3.1 (chapter 3.4).

3.1 Underground exposures

In the Hattorf mining field the intrusive rocks occur nearly exclusively in the form of dikes with widths generally varying between a few centimeters and about one meter. Wider dikes or bodies of basalt were observed comparatively rarely in the evaporite rocks (e.g., KNIPPING 1984; KNIPPING & HERRMANN 1985; KNIPPING 1989).

Since the thicknesses of the various dikes are not constant along their NS extension, the basaltic rocks occur in differing forms. For example, in narrow dikes or dike zones the portion of basalt occurring as round fragments is small compared with rock salt. These fragments are only a few centimeters in diameter. The brecciated basalt is, in part, completely decomposed into a light, earthy, friable substance (e.g., outcrop 19-ut). In contrast, the basalt in several of the wider dikes is hard, dark, and apparently fresh (fig. 3.1). The subsurface basalts consist predominantly of a gray to grayish-green rock exhibiting fissility or more rarely a friable groundmass. These rocks always contain fine, horizontal and vertical cracks which chiefly contain anhydrite, halite, and Fe-Ca-Mg carbonates.

The magmas generally penetrated the rock salt beds vertically, and the contact between basaltic and salt rocks is sharp (fig. 3.2). Macroscopically, the color and composition of the basalt groundmass appears to be uniform across the entire widths of the dike. Where visible, the contact zone to the rock salt is limited to an up to 2-cm-wide, yellowish mineral phase of ankeritic to ferrodolomitic carbonates and anhydrite.

In contrast to the sharp contact basalt-rock salt the basalt dikes in the K-Mg rocks of K1Th, K1H and mainly the Begleitflöze of K1H frequently have widths in the meter range and show an indistinct contact to the host rock (fig. 3.3). Excepting the

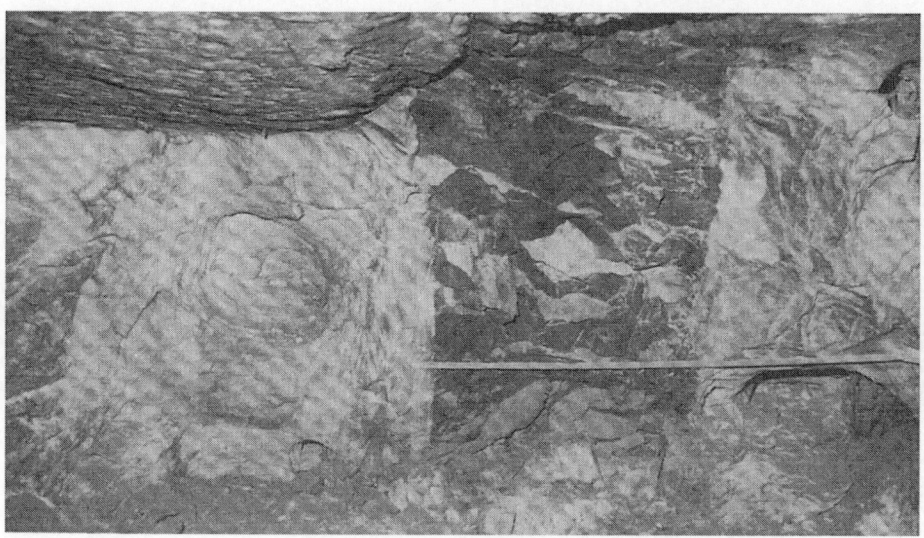

Fig. 3.1 Outcrop 1-ug in K1H. The massive, 6.5-cm-wide basanite dike exhibits comparatively few fissures and cracks. The light-colored patches in the dike are fracture surfaces covered by salt dust. Smaller gas cavities are seen on both sides of the dike.

Fig. 3.2 Outcrop 20-ug in Na1β (KNIPPING 1984). The 30-cm-wide dike exhibits fine, vertical and horizontal cracks filled mainly with anhydrite and halite, in addition to minor carbonate. The contact zone, up to 2 cm wide, consists of anhydrite, kieserite, and ferrodolomitic to ankeritic carbonates. The rock salt beds have clearly been vertically displaced as evidenced by the darker layers.

Fig. 3.3 Outcrop 17-ug in K1H. The spreading out of the greatly altered, friable olivine nephelinite in the potash horizons of K1H and the indistinct contact to the host rock are obvious.

carnallitite in K1Th, macroscopic features are rarely observed at the contact to the K-Mg rocks. Reddish polyhalite at the basalt-rock salt and sylvinite contact are noteworthy as well. Gas cavities are frequently found near such contact zones.

The groundmass of the apparently fresher basalts, in which olivine aggregates up to 1 cm in size are sometimes found, is fine-grained and dense (e.g., dike <E>). The basalt of dike <N> contains hornblende crystals, partially over 1.5 cm long, which are concentrated towards the center of the dike. Hornblende-bearing basalts of this type have not yet been observed anywhere else in the Northern Hessian Depression.

Native elemental sulfur occurs on the contact surfaces between dike and the kieseritic Hartsalz of K1H (sample 15/34) and especially in the western parts of K1H (dikes <N> and <O>); it is also found sporadically in the basalt itself. In addition, native sulfur occurs locally in greater quantities up to 40 m away from the basalt in the Hessen (K1H) potash salt seam. The portions of sulfur in the total rock vary from fine impregnations in the evaporites up to accumulations of nearly pure sulfur (figs. 3.4, 3.5).

The observations made at underground outcrops in the tunnels and at active workings are confirmed by the drill cores from various horizontal drillings. In these

Fig. 3.4 Basalt outcrop in district 4N, W46/110 (dike <O> west of outcrop 8-ug). The relationship between the occurrence of native sulfur and the basalt intrusion can be clearly seen next to the gas cavities and the evaporite layers which have been dragged upward.

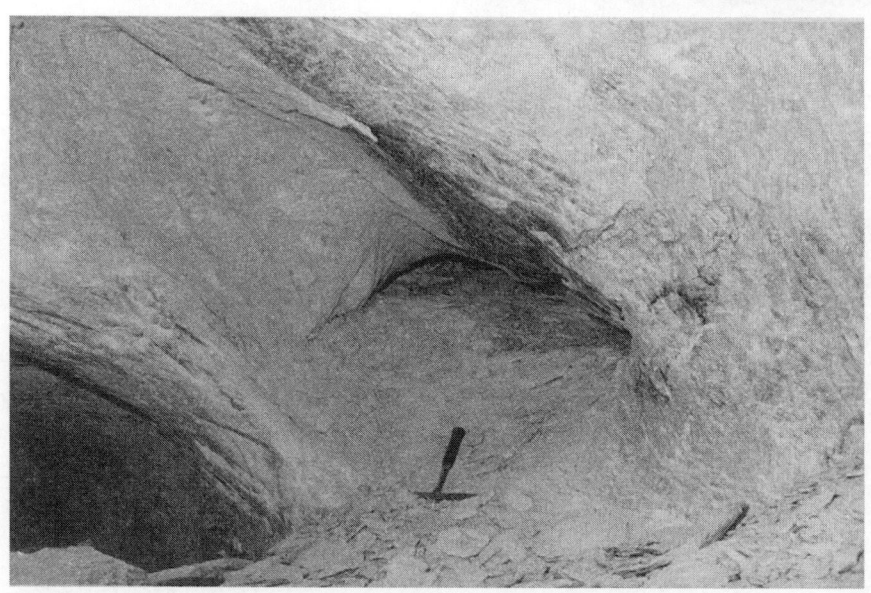

Fig. 3.5 Outcrop 61-ug in K1H. Very sulfur-rich salt rock thrown out by gas explosion (predominantly CO_2) 20 m away from the basalt dike. The original evaporite rock has locally been replaced completely by sulfur. The typical fissility of the gas-bearing salt rock can be seen.

cases, however, reliable data on the widths of the dikes is hard to obtain because the basalts were rarely encountered perpendicular to their strike and dip.

3.2 Surface exposures

The Hattorf mining field extends over the geological maps of sheet Friedewald (TK 25: 5125) and sheet Geisa (TK 25: 5225). All basalts striking across the mining field at the surface were sampled. Previous data on sheet Friedewald was available from BÜCKING (1921), and on sheet Geisa from a newer edition of LAEMMLEN (1975). Unfortunately, several of the basalt outcrops given in BÜCKING (1921) are no longer exposed due to overgrowth and weathering or are no longer accessible due to the East-West German border. Examples are the outcrops east of the road between Wehrshausen and Oberlengsfeld (350.5 m a.s.l.), north of Oberbreitzbach (Hohenroda farm; 353 m a.s.l.), and of the Eselskuppe on the bank of the Ulster between Philippsthal and Unterbreitzbach.

At the outcrops sampled it was difficult to get at the solid, unweathered basalt, excepting Soisberg (42-sf) and Trumbachsköpfchen (44-sf). The ca. 0.28 km² occurrence at Soisberg is the largest in the study area. Here, there are several crags of unweathered basalt containing olivine inclusions and showing distinct *Sonnenbrenner* effects at the summit. At Trumbachsköpfchen the remains of a small volcanic neck about 40 m in diameter and surrounded by tuff deposits several meters thick are exposed. Idiomorphic Hornblende crystals have been eroded out of these tuffs. The magma extruded here was obviously the same which formed dike <N>. Of the samples taken at the surface the least weathered basalt seemed to be that at Landeck castle, NE Oberlengsfeld.

In contrast, the rock in the abandoned quarry at Dreienberg south of Friedewald was intensively weathered. KÄDING (1962, p. 38) described an initial magma which was subsequently crosscut by three younger, NS-striking dikes. This multiple intrusion is no longer exposed today.

3.3 Sampling

Sampling underground and at the surface in the area of the Hattorf mine was done from 1983 to 1986. The following number of samples was taken:

	number of dikes	number of outcrops	number of basalt samples	samples f. isotope determination sulfur	carbon
subsurface		48	43	26	3
	15				
surface		7	8	-	-

16 of the 48 subsurface exposures were cores from horizontal drillings done by Kali & Salz AG from 1982 to 1985.

All samples taken at the active face were hand specimens. The freshest possible samples were taken underground for petrographically classifying the basalts. When the widths of the dikes allowed (> 50 cm), the samples were taken from the middle of the dike.

The individual basalt samples weighed up to 50 kg depending on the quality of the outcrops. About 1 kg of each evaporite rock was sampled for studying the sulfur and carbon isotopes. Samples were cleaned and macroscopically altered portions were removed directly at the outcrop.

After labeling, the basalt and evaporite samples were sealed in plastic bags to protect against moisture and contamination. This procedure was necessary for the basalts because the evaporite minerals in the small cracks and fractures were to be identified under the microscope. The same procedure was used for the basalt drill cores. About 1-5 kg was sampled depending on the length of the halved cores.

In addition, samples of compact langbeinite, kieserite, and polyhalite near the basalt dikes were taken. The δ^{34}S-values of these samples were determined for evidencing the possible genetic relationships between basalt intrusion and the formation of specific evaporite minerals.

Weathered material was also removed in the field from surface samples. The surface basalts generally appeared to be fresher than those from underground. Yet, at outcrops 43-sf (366 m a.s.l., southwest of the Hohenroda farm) and 45-sf (Lieshauk near Mansbach) all samples taken were slightly weathered.

3.4 List of samples and localities

Table 3.1

Explanations for the columns:

1. *ug* subsurface (underground) exposure
   sf surface exposure

   <A> to <P> numeration of dikes (cf. fig. 2.2). Dike <B> was not accessible for sampling.
   * Supplementary samples from KNIPPING (1984) and KNIPPING & HERRMANN (1985)

2. Samples for sulfur and carbon isotope-determinations are also labeled with *s* and *c*, respectively.

3. Abbreviations for the basaltic rocks

   ON olivine nephelinite
   LI limburgite
   BA basanite
   PT phonolitic tephrite
   nc not classified. Due to the intensive alteration the sample was not able to be classified according to the modern nomenclature for magmatic rocks.

4. Stratigraphic abbreviations (left of the virgule)

   Na1γ oberes (upper) Werra rock salt
   K1H Hessen potash seam
   Na1β mittleres (middle) Werra rock salt
   K1Th Thüringen potash seam

   Mineralogical abbreviations (right of the virgule)

   KC carnallitite
   KH Hartsalz
   Ksy sylvinite
   Na rock salt

5. The mining district (dist.) and, in part, the tunnel number (e.g., W49/111) are given for the underground exposures. The data on drill core samples consists of the district of the horizontal-drilling site, the number of the drilling, and the distance between the sample and the drilling site in meters. Unless given otherwise, the underground exposures are all located in horizon 1 of the Hattorf mine (K1H horizon). The geographical direction is given for the surface exposures.

   s southern tunnel wall
   n northern tunnel wall
   HD horizontal drilling
   VD vertical deep drilling

6. Thickness of the basalt dikes underground at the level of sampling. The thickness of the basalt dikes in which the drill cores were taken cannot be given (*n.d.*, no data) because the drilling was usually oblique to the strike and dip of the dikes.

1	2	3	4	5	6	7	8
exposure/ dike	sample no.	type of sample	stratigraphic horizon	sampling site	thickness of dike [cm]	date of sampling	notes
1-ug/<H>	1/2	BA	K1H/KH	Westfeld entrance via 3rd south tunnel, s	65	9/24/1984	-
2-ug/<F>	2/5	BA	K1H/KH	Westfeld entrance, 5th south tunnel, n	270	9/24/1984	widening of basalt dike in K1H
3-ug/<H>	3/7	BA	Na1γ/Na	about 450 m west of 2-ug	45	9/24/1984	-
4-ug/<G>	4/11	ON	K1H/KH	same as 3-ug	26	9/24/1984	-
5-ug/<M>	5/13	BA	Na1γ/Na	dist. 4 entrance, 100 m south of point HD 109, n	23	9/24/1984	basalt dike displaced 2.5 m
6-ug/<M>	6/14	BA	Na1γ/Na	7 m west of 5-ug	19	9/24/1984	-
7-ug/<M>	7/15 7/115[s]	BA langbeinite	Na1γ/Na	40 m south of 5-ug and 6-ug	37	9/24/1984 5/27/1986	40 m west of dike dense langbeinite
8-ug/<N>	8/16[s] 8/17 8/18[s] 8/88[s] 8/113[s] 8/114[s] 8/114[c] 8/130[s] 8/133[s]	sulfur BA sulfur sulfur sulfur sulfur CO_2 sulfur sulfur	K1H/KH	dist. 4W entrance, G47, n	80	9/24/1984 9/24/1984 9/24/1984 9/5/1985 9/27/1985 9/5/1985 9/5/1985 9/5/1985 9/27/1985	8/16 and 8/18 native sulfur at the basalt contact; 8/88 native sulfur in evaporite rock about 40 m west of basalt dike; 8/113 and 8/133 sulfate for isotope determination, free of native sulfur; 8/114 gas-bearing salt (Knistersalz) from a cavity 10 m west of the basalt dike; 8/130 salt rock impregnated with sulfur
9-ug/<N>	9/19[s]	sulfur	K1H/KH	dist. 4W, W49/107	-	9/24/1984	native sulfur with kieserite, anhydrite, and pyrite at a basalt contact; only the western contact is exposed
10-ug/<E>	10/22	ON	Na1γ/Na	west tunnel, 1 100 m west of Ransbach shaft, n	5	9/25/1984	gas cavity field partially with a little basaltic material
12-ug/<O>	12/25	PT	K1H/KH	dist. 4W, W49/111, s	115	9/25/1984	basalt interspersed with halite crystals

Tab. 3.1 List of samples and localities, continued

1	2	3	4	5	6	7	8
exposure/ dike	sample no.	type of sample	stratigraphic horizon	sampling site	thickness of dike [cm]	date of sampling	notes
13-ug/<N>	13/27 13/28^s 13/30	BA polyhalite BA	Na1γ/Na	dist. 5NW, W23/105, n 5 m east of 13/27, s	55 30	9/25/1984 9/25/1984 9/25/1984	polyhalite at basalt contact; basalt dike displaced by about 5 m
15-ug/<D>	15/33a 15/33b 15/34^s	BA BA sulfur	K1H/KH	Südfeld entrance, 150 m north of point HD 100, s	110	11/7/1983 11/7/1983 9/26/1984	basalt dike widens in K1H; 15/34^s with small amounts of native sulfur at the contact
16-ug/<I>	16/37 16/81^s 16/82^s	LI sulfur sulfur	Na1β/Na	dist. 2NW, W100/68, s	40	9/26/1984 9/26/1984 9/26/1984	dike widens in K1H; 16/81^s and 16/82^s from kieserite bed (99 % ks) about 1 m east of contact
17-ug/<I>	17/39	LI	K1H/KH	10 m east of 16-ug, n	25	9/26/1984	connected with 16-ug by a sill
18-ug/<G>	18/40	ON	K1H/KH	dist 3NE, W85/31, 650 m northeast of VD Mansbach II, s	23	9/26/1984	basalt dike widens in K1H
19-ug/<I>	19/41	nc	Na1β/Na	south. dist. 1S entrance, n	4	9/26/1984	greatly decomposed; sample < 0.5 kg
20-ug/<A>	20/43	ON	Na1β/Na	tunnel to Wintershall, n	55	9/27/1984	strike of this dike differs from those of other basalt dikes in the mining field (fig. 2.2)
21-ug/<C>	21/46	ON	K1H/KH	about in the middle of the southwest tunnel, s	100	9/27/1984	
22-ug/<E>	22/48	ON	K1Th/KH	2nd horizon, dist. 0W, W65/23-24, n	22	9/27/1984	dike widens in the upper part of K1Th
23-ug/<N>	23/51^s 23/52 23/52^s 23/53^s	sulfur BA sulfur sulfur	Na1γ/Na	dist. 4SW, HD 129/82, depth of 109.2 m	n.d.	9/27/1984 9/27/1984 9/27/1984 9/27/1984	23/51^s native sulfur at basalt contact; 23/52^s sulfur in the basalt matrix

1	2	3	4	5	6	7	8
exposure/ dike	sample no.	type of sample	stratigraphic horizon	sampling site	thickness of dike [cm]	date of sampling	notes
24-ug/<N>	24/55, 24/54ˢ	BA, sulfur	Na1γ/Na	dist. 4SW, HD 129/82, depth of 111.1 m	n.d.	9/27/1984, 9/27/1984	native sulfur at basalt contact
27-ug/<N>	27/59	BA	Na1γ/Na	dist. 4SW, HD 129/82, depth of 137.8 m	n.d.	9/27/1984	-
28-ug/<N>	28/61, 28/62ˢ	BA, sulfur	K1H/KH	dist. 4SW, HD 130/82, depth of 52.7 m	n.d.	9/27/1984, 9/27/1984	native sulfur at basalt contact and in the basalt
29-ug/<N>	29/63	BA	Na1γ/Na	dist. 4SW, HD 130/82, depth of 163.2 m	n.d.	9/27/1984	-
30-ug/<N>	30/65	BA	Na1γ/Na	dist. 4W, HD 130/82, depth of 172.4 m	n.d.	9/27/1984	-
31-ug/<O>	31/68	PT	Na1γ/Na	dist. 4W, HD 130/82, depth of 235.4 m	n.d.	9/27/1984	-
32-ug/<N>	32/70	BA	Na1β/Na	dist. 5W, HD 131/83, depth of 96.8 m	n.d.	9/27/1984	-
33-ug/<O>	33/723	PT	Na1β/Na	dist. 5W, HD 131/83, depth of 209.4 m	n.d.	9/27/1984	-
34-ug/<J>	34/75	LI	Na1γ/Na	dist. 2NW, HD 132/83, depth of 377.5 m	n.d.	9/27/1984	drilling ended in the basalt
35-ug/<J>	32/70	LI	Na1γ/Na	dist. 2NW, HD 134/83, depth of 358.0 m	n.d.	9/27/1984	-
36-ug/<J>	36/78	LI	K1H/KH	dist. 2NW, HD 135/84, depth of 119.0 m	n.d.	9/27/1984	-
37-ug/<K>	32/79	LI	Na1γ/Na	dist. 2NW, HD 135/84, depth of 430.5 m	n.d.	9/27/1984	-
38-ug/<O>	38/831ˢ, 38/832ˢ	sulfur, sulfur	K1H/KH	dist. 5W, W26/108, n; dist. 5W, W26/108, s	-	4/25/1985, 4/25/1985	native sulfur at basalt contact

Tab. 3.1 List of samples and localities, continued

1	2	3	4	5	6	7	8
exposure/ dike	sample no.	type of sample	stratigraphic horizon	sampling site	thickness of dike [cm]	date of sampling	notes
39-ug/<O>	39/84[s]	sulfur	K1H/KH	dist. 5W, W23/110	-	4/25/1985	native sulfur and pyrite in a joint between *Flocken-* and *Bändersalz* about 10 m away from the basalt
40-sf/<O>	40/21	PT	-	NE Oberlengsfeld, at the eastern wall of Landeck castle	-	9/24/1984	-
41-sf/<L>	41/90	ON	-	SW Wehrshausen, 447.5 m a.s.l.	-	6/1/1984	-
42-sf/<L>	42/91	ON	-	Hohenroda-Soislieden, SE slope of Soisberg	-	6/1/1984	*Sonnenbrenner*
43-sf/<H>	43/92	BA	-	SW Hohenroda farm near Oberbreitzbach, 366 m a.s.l.	-	6/1/1984	weathered
44-sf/<N>	44/93	BA	-	road from Ransbach to Schenklengsfeld, junction Wehrshausen, Trumbachsköpfchen	-	11/7/1983	tuffs are exposed next to the basalt
45-sf/<D>	45/94	BA	-	NW Siffig settlement, near Mansbach, Lieshauk	-	11/7/1983	weathered
46-sf/<P>	46/86 46/87	nc PT	-	SSW Friedewald, Dreienberg, abandoned quarry behind rifle-club house	-	9/4/1985 9/4/1985	poor exposure due to vegetation and the shooting range in quarry
50-ug/<O>	50/85	PT	Na1β/Na	dist. 5W, W24/110, n	180	9/4/1985	traces of native sulfur at the basalt
51-ug/<G>	51/89	ON	K1H/KH	dist. 3NE, W82/31 - 32	22	9/5/1985	-

1	2	3	4	5	6	7	8
exposure/ dike	sample no.	type of sample	stratigraphic horizon	sampling site	thickness of dike [cm]	date of sampling	notes
52-ug/<H>	52/95	BA	K1H/KH	Südfeld entrance via 6th south tunnel, about 600 m south of 3-ug and 4-ug, s	23	9/5/1985	-
53-ug/<H>	53/96	BA	K1H/KH	5.5 m east of 52-ug, s	42	9/5/1985	-
54-ug/<L>	54/97	ON	K1H/KH	dist. 5, s	20	9/6/1985	-
55-ug/<K>	55/98	LI	Na1γ/Na	dist. 3W, HD 140/85, depth of 684.5 m	n.d.	9/6/1985	-
56-ug/<K>	56/99	LI	Na1γ/Na	dist. 3W, HD 140/85, depth of 636.5 m	n.d.	9/6/1985	native sulfur at basalt contact and in the basalt
57-ug/<J>	57/140	LI	Na1γ/Na	dist. 3W, HD 140/85, depth of 254 m	n.d.	9/6/1985	-
58-ug/<K>	58/141	LI	Na1γ/Na	dist. 2NW, W101/74-75, n	10	9/27/1985	-
59-ug/<O>	59/144[s]	sulfur	K1H/KH	dist. 5W, W23/110	-	9/6/1985	massive sulfur in a gas cavity (*Knistersalz*)
60-ug/<O>	60/142[c] 60/142[c] 60/142[s]	sulfur CO_2 langbeinite	K1H/KH	dist. 5W, W21/110 - 111	-	4/18/1986 4/18/1986 4/18/1986	massive sulfur ($> 10^2$ kg) in K1H and in the 1st Begleitflöz (*Knistersalz*); little langbeinite about 20 cm away from a basalt
61-ug/<O>	61/143[s] 61/143[c]	sulfur CO_2	K1H/KH	dist. 5W, W20/110 - 111	-	4/18/1986 4/18/1986	massive sulfur in a gas cavity 30 - 40 m west of a basalt (*Knistersalz*)
62-ug/<J>	62/146[s]	sulfate	K1H/KH	old entrance dist. 4S, n	-	4/18/1986	sylvinite bed in the *Bändersalz* 20 m east of a basalt dike
1D-ug*/<E>	1D/12	ON	K1Th/KC	2nd horizon, W1/8	-	10/28/1982	isolated, ellipsoidal body of basalt in the *Trümmer* carnallitite (visible dimensions of 45 · 100 cm)
2D-ug*/<I>	2D/24	nc	Na1β/Na	tunnel to dist. 3S	30	10/28/1982	-

4 Sample preparation and analytical methods

4.1 Preparation of basaltic samples for mineralogical and chemical study

The surface and subsurface basalt samples were prepared for mineralogical and chemical study according to the scheme in fig. 4.1. The steps labeled (1) and (2) are explained as follows:

1. All of the subsurface samples contained varying amounts of water-soluble, evaporite minerals (e.g., halite, anhydrite) in fine cracks and fissures, in vesicular cavities, and finely distributed in the matrix of the rock. Therefore, very thin mineral oil (e.g., Shell KS 111 or MS 4919) was used instead of water for preparing the thin sections (sawing and polishing).

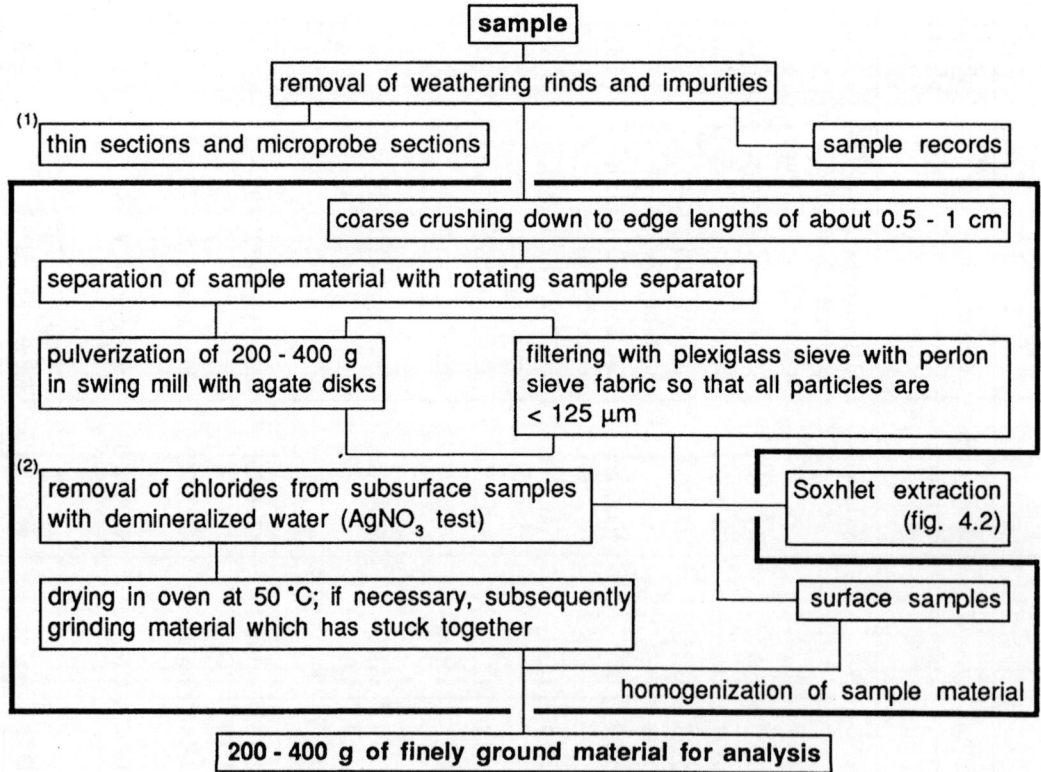

Fig. 4.1 Procedure for preparing basalt samples. Explanations for numbered steps in chapter 4.1.

2. FRANKE (1974) suggested that samples be washed for about 10 days. Based on the author's experience, chloride minerals are quantitatively removed after washing the samples for one day using a paper filter (e.g., Schleicher & Schüll, Schwarzband) and suction apparatus (KNIPPING 1984). In doing so, anhydrite will also be removed partially or totally from the sample material depending upon the volumes and the NaCl concentration of the washing solution.

4.2 Preparation for sulfur-isotope determinations

Sulfur isotopes were determined in the following rock types: (1) Evaporites that contained macroscopic native sulfur in addition to sulfates (kieserite, anhydrite) and were sampled directly at the basalt contact as well as up to 40 m away from the basalt. (2) Compact langbeinite, kieserite, and polyhalite near the basalt dikes and at the contacts to them. (3) Basaltic rock from various dike systems, of which only sample 23/52 contained macroscopically larger portions of native sulfur in the form of crystal aggregates. (4) Fine pyrite crystals which had grown on the contact surface to the basalt.

The procedure for sample preparation corresponds, for the most part, to the proven method used for years in the isotope lab at the Institute of Geochemistry of the University of Göttingen.

Preparation begins with the procedure described in step 2 below, when the sample contains no native sulfur (see fig. 4.2). Steps (1) - (4) in fig. 4.2 are explained as follows:

1. About 5 g of sulfur-bearing evaporite samples are finely ground in an agate mortar. Depending on amount of sulfur contained a portion of the sample is heated to about 60 °C in a porcelain dish on a sand bath after adding about 150 ml CCl_4 (ventilation!) until the native sulfur has completely dissolved (boiling point of CCl_4 = 78.5 °C; transition between rhombic and monoclinic sulfur at 95.5 °C; data on solubility of rhombic sulfur in CCl_4 in, e.g., D'ANS et al 1967). The sulfur-bearing CCl_4 is separated from the undissolved minerals using a folded filter (see 2 for treatment of residue).

   The filtrate is dried on the sand bath at 60 °C until the sulfur crystallizes. Then, 3.4 mg of the dried filtrate and 80 mg V_2O_5 are placed in a ca. 15-cm-long SiO_2-glass tube. The sample/V_2O_5 ratio must be observed so that the formation of SO_2 (see below) occurs quantitatively, and other sulfur oxides (SO_3 and SO) are not formed.

   For converting sulfur into SO_2 with V_2O_5 the SiO_2-glass tube is evacuated, sealed, and placed in an electric oven at 1 000 °C for 1.5 minutes. After cooling the SO_2 can be measured with a mass spectrometer.

2. The sulfate-bearing residue left after filtration (see also 1) or the sample of evaporite rock containing no native sulfur is mixed in a beaker with 500 ml of distilled water on a magnetic stirrer while simultaneously heating to 80 °C - 90 °C. Most sulfate minerals (anhydrite, kieserite, polyhalite, langbeinite) dis-

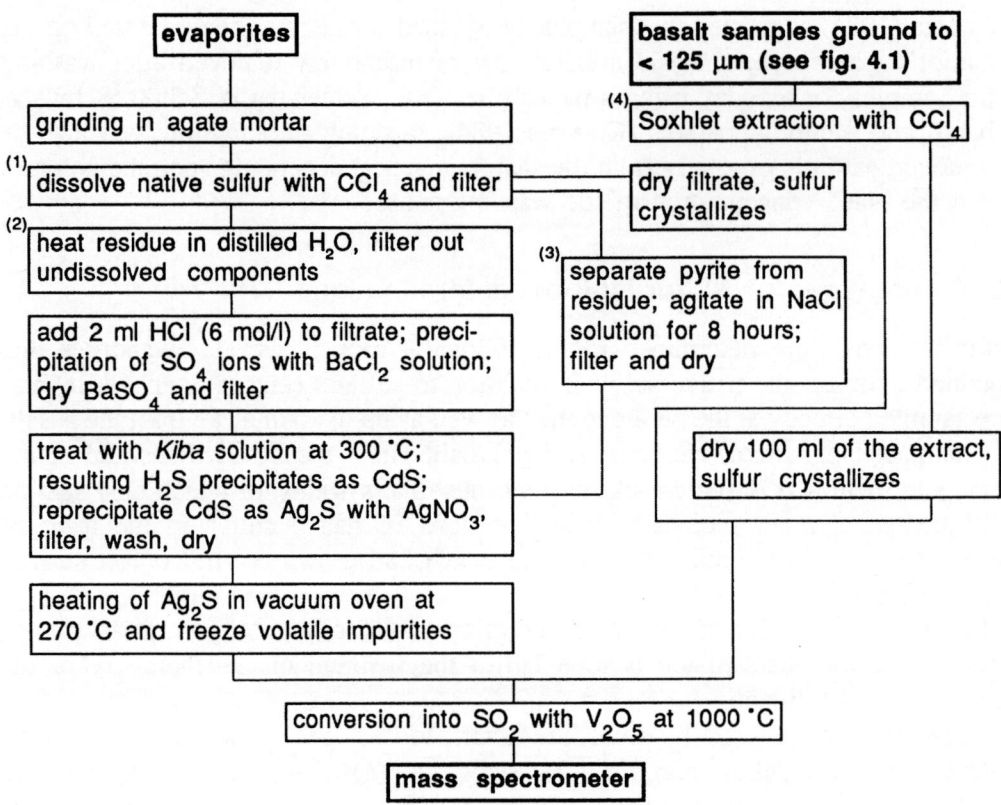

Fig. 4.2 Procedure for preparing evaporite and basalt samples for sulfur-isotope determination. Explanation to steps 1 - 4 in chapter 4.2.

solve in about 1 - 2 hours. The solutions are filtered with filter paper. After adding 2 ml of hydrochloric acid (HCl/H$_2$O volume ratio of 1/1, equaling 6 mol HCl/l; see JANDER & BLASIUS 1979, p. 181) the sulfate is precipitated as BaSO$_4$ by adding an excess amount of BaCl$_2$ solution (10 g in 100 ml H$_2$O) when the filtrate has been brought to just under the boiling point. After 24 hours the BaSO$_4$ precipitate is filtered using membrane filters and a vacuum pump, washed in distilled H$_2$O, and dried for 10 hours at 50 °C. Then, 70 mg of the BaSO$_4$ is weighed out and treated with the so-called *Kiba* reagent (80 g SnCl$_2$ in 1000 g H$_3$PO$_4$) for 30 minutes at 300 °C. The H$_2$S forming from the sulfate is passed through N$_2$ into a Cd acetate solution and precipitates as CdS. The CdS is reprecipitated as AgS$_2$ with a 2 % AgNO$_3$ solution, filtered and washed over a membrane filter. The resulting AgS$_2$ is dried at 50 °C. Then, 34 mg of silver sulfide and 80 mg of V$_2$O$_5$ are weighed out and placed in two separate SiO$_2$-glass tubes. The volatile impurities are driven off in a vacuum oven at 250 °C. The 34 mg of sulfide and 80 mg of V$_2$O$_5$ are subsequently mixed together in a SiO$_2$-glass tube which is evacuated at 100 °C and sealed off. Further preparation proceeds analogously to step 1.

3. The residue obtained by filtering the CCl_4 (see point 1) often contains pyrite in addition to sulfates anhydrite, kieserite, and polyhalite. The amount of pyrite in sample 9/19 was sufficient for isotope determination. The pyrite was sorted out under the binocular microscope. The remaining portions of sulfate minerals (primarily anhydrite) were removed by agitating them with the pyrite in a NaCl solution for about 8 hours. The NaCl solution consists of 20 g NaCl in 100 g H_2O (20 °C) and can disolve 0.823 g $CaSO_4$ (D'Ans 1933). After filtering with a quantitative paper filter (e.g., Schleicher & Schüll, Schwarzband) the pyrite was also able to be removed with *Kiba* solution and treated further like the $BaSO_4$ (see step 1).

4. Isolating sulfur from the groundmass of the subsurface basalts involved extracting 200 g of the chloride-bearing, finely ground basalt sample in a Soxhlet apparatus with about 700 ml of CCl_4 over a period of 24 hours. Following the subsequent redistillation of 600 ml of the solvent the native sulfur crystallized out of the remaining 100 ml by evaporation in a beaker (ventilation!). The basalt from outcrop 23 was the only one bearing macroscopically larger crystals of sulfur (up to 2 mm in diameter). About 1 g of sulfur was obtained from this basalt. The other subsurface samples yielded 1 - 10 mg of a sulfur-yellow, oily, bituminous substance.

The highly viscous substance was quantitatively transformed by redissolving it in a porcelain dish with about 5 ml of CCl_4 and transferring it into a SiO_2-glass tube using a glass capillary tube. The CCl_4 was removed using water-jet suction and an intermediary Woulff flask. Weighing the SiO_2-glass tube before and after addition of the substance indicated that 5 - 10 mg of sulfur were obtained.

Since the subsurface basalts also contained larger amounts of sulfates (e.g. anhydrite, polyhalite) in addition to elemental sulfur and pyrite at the contact surfaces and in fine cracks and fractures, determining the isotope distribution of the total sulfur did not appear to be very practical. The method of dissolving sulfides with HCl according to Schneider (1970) was not employable for separating the sulfur bound as sulfide in the groundmass. Several subsurface samples contained finely disseminated pyrite at the contact. In reaction with HCl the decay of pyrite to H_2S is not quantitative. Fractionated, thermal decomposition of the sulfates and sulfides in oxygen was used as an alternative on 15 basalt samples (Brumsack 1981). However, the isotope determinations did not yield any reproducible values. Perhaps the separation was not quantitative due to the large amount of sulfate and small amount of sulfide.

4.3 Preparation for carbon-isotope determinations

During preparation of the samples for sulfur-isotope measurement it became evident that all evaporite samples bearing native sulfur are also very rich in gas (*Knistersalz*). According to Ackermann et al (1964) gas-bearing drill core samples of K1Th (carnallitite, sylvinite; Marx Engels mine, formerly Menzengraben, East Germany) contained up to 0.6 - 14.0 ml gas/100 g salt rock. An average of about 3.6 ml of gas was fixed in 100 g of salt rock. The gas inclusions are composed of 84 vol% CO_2 on the average. The carbon isotope-distributions of the CO_2 were determined for genetic interpretation. Here I would like to thank Dr. A. Kirchhoff and R. Przybilla for their stimulation and technical assistance.

Precipitating carbon as carbonate with $Ba(OH)_2$ and $BaCl_2$ in a buffer solution with a pH of 10 (HOEFS 1973; KOCH & VOGEL 1980) is not a suitable preparation method for determining isotope distributions. BAERTSCHI (1952) and more recently USDOWSKI & HOEFS (1986) were able to show that $BaCO_3$ is depleted in ^{13}C during precipitation. This kinetic fractionation produces a shift in the $\delta^{13}C$ values to lower values through the enrichment of ^{12}C.

The alternative was to measure directly the CO_2 enclosed in the salt rock. About 350 g of a gas-bearing salt rock sample was weighed out in larger pieces in a 1 000 ml flask (about 360 g of NaCl will dissolve in 1 000 g of H_2O at 25 °C, recalculated from D'ANS 1933). About 12 ml of CO_2 was obtained with a production ratio of 4 ml gas/100 g salt rock with around 80 vol% CO_2. The gas released was collected in the flask filled to the rim with doubly distilled water. The CO_2 concentration in the atmosphere (P_{CO_2} = 32 Pa = $10^{3.5}$ atm) is negligibly low when compared with the CO_2 concentration expected in the sample. It follows that even when the doubly distilled water is CO_2 saturated, contamination of the sample by atmospheric CO_2 ($\delta^{13}C$ around -7 per mil, HOEFS 1987) does not become a problem. Just after the flask is sealed with a ring and silicon seal, the first gas bubbles already begin to collect. The first 3 - 5 ml of gas was extracted through the silicon seal using a gas syringe. This method should prevent contamination of the sample by atmospheric CO_2 clinging to the sample.

After the water-soluble minerals in the sample dissolved completely, the gas which had collected under the seal of the flask was transferred with a steel capillary to an evacuated SiO_2-glass tube. The gas was then frozen with liquid nitrogen, and the SiO_2-glass tube was placed directly at the mass spectrometer for measurement after being sealed with an oxygen-hydrogen flame.

Another gas sample was taken from the flask after 2 - 6 days and frozen for a double check. In view of the simple preparation method the measured values correlated well (see chapter 6).

Analogous to the samples for sulfur-isotope determination the bitumen samples from the Soxhlet extraction (chapter 4.2) were converted into CO_2 in evacuated SiO_2-glass tubes with V_2O_5 at 1 000 °C for CO_2 measurement in the mass spectrometer.

4.4 Analytical methods

The following contains information on the methods used for determining the mineralogical and chemical composition of the basalt and evaporite samples.

Microscopy:

Basalt samples from the middle of a dike and from the contact to the salt rock were studied under the microscope. Part of the evaporite samples for sulfur-isotope measurement were also studied in this way. The mineralogical composition of the samples aided the petrographical classification of the rocks and their genetic interpretation.

Diffractometry:

Diffractometry supplemented microscopy by determining the mineralogical composition of the sulfur-bearing evaporites. Diffractometry was done with a Philips goniometer (PW 1130) with Cu X-ray tube under the following conditions:

voltage	50 kV
current	20 mA
counting rate	100 - 400 cps
integration constant	1s
angular velocity	1° 2θ/min

The X-ray data for evaporite minerals were compiled by BRAITSCH (1962, 1971: tab. 3).

Chemical analysis:

The major and minor elements determined in the basaltic rocks and the respectively method of analysis are as follows:

SiO_2	- titrimetry
TiO_2, Al_2O_3	- ICP-AES
Σ Fe as Fe_2O_3, MnO, MgO, CaO	- AAS, flame
FeO	- titrimetry
Na_2O, K_2O	- AES, flame
ΣH_2O	- gravimetry (Penfield)
P_2O_5	- spectrophotometry
ΣC as CO_2	- coulometry
Ni	- AAS, graphite tube

ICP-AES - inductively coupled plasma atomic emission spectrometry (instrument: ARL 35 000)
AAS   - atomic absorption spectrometry (instrument PE 4 000)
AES   - atomic emission spectrometry (instrument PE 4 000)

The samples for ICP-AES, AAS and AES measurements were dissolved in an autoclave with a mixture of hydrofluoric acid and perchloric acid and subsequently evaporated four times with hydrochloric acid on a hot plate. The analyses of inhouse reference samples of known composition (alkali-olivine basalt, picrite basalt) served as controls.

The chemical analysis has been described in detail by HERRMANN (1975) and HEINRICHS & HERRMANN (1989). An introduction to AAS, AES, and ICP-AES including the measuring procedure is given by WELZ (1976) and HEINRICHS & HERRMANN (1989). Coulometric sulfur measurements were done as prescribed by LANGE & BRUMSACK (1977).

Microprobe:
Evidence of the origin and genesis of the hornblende in the basanite from dike <N> should be obtained with the microprobe. Hence, some major and minor elements in hornblende crystals from five different outcrops were determined. Measurements were taken with a ARL electron emmission microprobe SEMQ using international standards for comparison.

X-ray fluorescence spectroscopy (XRF):
The following major and minor components were determined in the basalts: SiO_2, Al_2O_3, MgO, CaO, and MnO. $SrNO_3$ served as an internal standard. Samples were prepared as described by SCHNEIDER & SCHULZ-DOBRICK (1976). Measurements were taken with an ARL 72 000 X-ray spectrometer and evaluated with international reference samples.

Instrumental neutron activation analysis (INAA):
The following elements were determined in the basalts with INAA: Na, Sc, Cr, Fe, Co, La, Ce, Sm, Eu, Tb, Yb, Lu, and Hf. The samples were radiated with thermal neutrons in the experimental reactor of the Medical University of Hanover (duration of radiation 8 hours; neutron flux of about $2 \cdot 10^{12}$ n/cm^2sec. Gamma activity was measured with a Nuclear Data multichannel analysis system ND 6603 using USGS reference samples W-1 and BCR-1 for comparison. This method is described, for example, by JACOBS et al. (1977) and MUECKE (1980).

Mass spectrometry:
Sulfur and carbon isotopes were determined using a Finnigan MAT 251 mass spectrometer. The samples were first purified with liquid nitrogen and a mixture of liquid nitrogen and n-pentan before admission of the measuring gases.

To obtain evidence of possible gaseous impurities a portion of the mass spectrum was recorded before the actual measurements. Numerous, undeterminable, organic foreign gases were detected in the bitumen samples from the Soxhlet extraction.

Only in two samples the degree of contamination was so low that the measurements were able to be evaluated. However, the low CO_2 content in these samples necessitated a substantially smaller container on the mass spectrometer ('mini-gas-induction system'). The results are given with reference to international standards CDT ($\delta^{34}S = 0$ per mil) and PDB ($\delta^{13}C = 0$ per mil).

Method-specific standard deviations (absolute) were calculated for the various elements and concentration ranges. These standard deviations were obtained by repeated measurement (KAISER & GOTTSCHALK 1972). Equation 3.03 in DOERFFEL (1965, p. 27; see also HEINRICHS & HERRMANN 1989) was used for calculation. Table 4.1 shows the results for the approximate values s of the standard deviation σ. Two different methods of analysis were used for determining several components. The mean value of the results of the various methods is then calculated for analyzing the chemical components in the basalt samples.

Tab. 4.1 Absolute error of the methods of analysis employed (approximate value s for the standard deviation σ). The partially wide variation in the chemical components of the various basalts necessitated the calculation of s for the differing ranges of concentration.

ICP-AES - inductively coupled plasma atomic emission spectrometry
AAS   - atomic absorption spectrometry
AES   - atomic emission spectrometry
INAA   - instrumental neutron activation analysis
XRF   - X-ray fluorescence spectroscopy

major and minor components, wt%			
components	method	range of concentration from - to	$s_{absolute}$
SiO_2	titrimetry	35 - 55	± 0.4
	XRF		± 0.4
TiO_2	ICP-AES	1 - 4	± 0.1
Al_2O_3	XRF	10 - 20	± 0.2
	ICP-AES		± 0.2
Σ Fe as Fe_2O_3	AAS	1 - 8	± 0.2
	INAA		± 0.1
FeO	titrimetry	2 - 8	± 0.1
MnO	AAS	0.1 - 0.3	± 0.01
	XRF		± 0.01
MgO	AAS	2 - 13	± 0.1
	XRF		± 0.1
CaO	AAS	2 - 12	± 0.3
	XRF		± 0.3
Na_2O	AES	0.3 - 6	± 0.1
	INAA		± 0.1
K_2O	AES	0.8 - 5	± 0.1
ΣH_2O	gravimetry	2 - 6	± 0.1
P_2O_5	spektrophotometry	0.3 - 1	± 0.05
Σ C as CO_2	coulometry	0.1 - 4	± 0.04

Tab. 4.1 Absolute error of the methods of analysis, continued

trace elements, μg element/g rock (ppm)			
components	method	range of concentration from - to	$s_{absolute}$
S	coulometry	50 - 4 000 4 000 - 22 000	± 40 ± 200
Sc	INAA	3 - 40	± 0.4
Cr	INAA	10 - 100 100 - 600	± 6 ± 20
Co	INAA	1 - 10 10 - 60	± 0.5 ± 1
Ni	AAS	10 - 100 100 - 350	± 6 ± 10
La	INAA	30 - 130	± 3
Ce	INAA	50 - 250	± 9
Sm	INAA	7 - 22	± 0.7
Eu	INAA	1 - 7	± 0.1
Tb	INAA	0.6 - 3	± 0.3
Yb	INAA	1 - 4	± 0.1
Lu	INAA	0.2 - 0.5	± 0.06
Hf	INAA	3 - 18	± 2

5 Composition of the basaltic rocks

Mineralogical and chemical studies evidence that sections of the individual basalt dikes which seem to correlate geographically are actually the products of two different intrusive events in some cases. For example, in fig. 2.2 it appears that dike <G> might be the southern extension of basanite dike <H>. Dike <G>, however, is an olivine nephelinite. There are also other cases in which portions of basalt dikes separated by great distances and also lacking any direct connection between the individual sections can be assigned to the same dike or intrusive event based on their identical petrographies (samples 54-ug, 41-sf, and 42-sf, dike <L>). Since, in part, only sporadic samples were able to be taken over the NS extension of some dikes (e.g., dike <E>), the numeration selected for this study is not final and only serves as preliminary orientation.

Due to the specific changes in and particularities of the subsurface basalts the conventional classification of volcanic rocks based on the chemical composition and the normative mineralogy (CIPW norm) calculated with the chemistry can only be applied to a few samples (see chapter 5.2). Hence, microscopy was primarily employed for petrographically classifying the various intrusive rocks according to the recent nomenclature of basaltic rocks (chapter 5.1). In addition the freshest possible sample was taken from each dike. In total about 100 basalt thin sections were studied. The quantitative mineralogical composition was not able to be determined under the microscope due to the fine-grained or glassy character of most samples.

5.1 Petrography

The subsurface basalts contain greater amounts of H_2O and CO_2 typical for anchibasaltic to ultrabasic dike rocks due to the influence of OH-bearing minerals (phyllosilicates) and carbonates (see WIMMENAUER 1973). Nonetheless, in this study they are referred to as basaltic rocks in the true sense since they intruded up into relatively shallow levels as well as extruded onto the surface.

The rocks occurring in the field area are olivine nephelinites (ON), basanites (BA), limburgites ('hyalobasanite', LI), and phonolitic tephrite (PT). Samples from dike <I> and sample 46/86 (dike <P>) were not able to be classified due to intense autometamorphism (nc = not classified). Further subdivision into other varieties (e.g., hornblende basanite, nepheline basanite, basanitic alkali-olivine basalt) was not attempted because of the gradual transitions between the individual basalt types found in the working area. The rock of dike <O> classified as a phonolitic tephrite is of particular interest being a product of differentiation. This is seen in the mode of occurrence of the minerals and the degree of variation in the mineralogical and chemical composition within the dike.

The following mineralogical criteria were used for describing the intrusive rocks (e.g., WEDEPOHL 1983; WIMMENAUER 1985):

olivine nephelinite - melanocratic, only one pyroxene, foids, no plagioclase,

limburgite (hyalobasanite) - only one pyroxene, abundant microlitic glass; foids and plagioclase can not always be detected microscopically,

basanite - melanocratic, only one pyroxene, olivine, plagioclase, foids, > 5 % normative nepheline,

phonolitic tephrite - like basanite but with kalifeldspar (< plagioclase); tephrite (in the strictest sense) or leucotephrite with < 35 % mafic minerals; commonly lacking olivine.

The petrography of the intrusive rocks discussed in the following and the occurring particularities are compiled in table 5.1 in order to compare the various dikes more easily. The studied rocks generally have a porphyritic texture. Deviations from this are also noted.

Olivine

Xenomorphic olivine occurs as aggregates in several basaltic rocks (dikes <E> and <N>). The largest of the individual aggregates does not exceed 3 mm in diameter. The olivine displays translation lamellae typical of peridotite olivine. In addition to olivine, pale-green clinopyroxene and light-brown orthopyroxene, which are marked by their high relief, and dark-olive-brown, nearly opaque spinel, are also observed in these relatively small peridotite xenoliths (up to about 2.5 cm Ø). These fragments of mantle rock are possibly harzburgite or lherzolite which are typical for the basalts of the Northern Hessian Depression (e.g., OEHM et al. 1983; HARTMANN 1986). A more exact classification was not possible due to the limited size and number of inclusions.

Short-prismatic-idiomorphic olivine phenocrysts were much more common in the groundmass of the basaltic rocks. These crystals are always smaller than 2 mm. This mineral is frequently decomposed to a fine-fibrous, pale-yellowish-green, weakly pleochroitic substance resulting in typically pitted, fine cracks and displacements. The product of this decomposition is (Fe(II)-bearing?) chrysotile. The microcrystalline mineral completely replaces olivine preferably in the more glassy portions of the rock.

In one sample from outcrop 46-sf (dike <P>) the perimeters of the olivine crystals are made up of a reddish-yellow substance. This so-called 'iddingsite' only occurs as pseudomorphs after olivine. Microscopically, iddingsite has all the properties of a homogeneous mineral. However, it is composed predominantly of goethite, in part with hematite, and submicroscopic clay minerals (TRÖGER 1967).

Idiomorphic olivine was also a component in the groundmass of several of the dikes (e.g., dikes <C>, <G>, and <M>). These much smaller crystals (< 1 mm Ø) have also been converted partially into chrysotile.

Pyroxene, amphibole

Clinopyroxene is another frequent phenocryst in the basaltic rocks, occurring nearly always as zoned idiomorphic augites which are frequently corroded. The margins are

Table 5.1 Petrography of the basalts

Explanations for the columns:

1. <A> to <P> numeration of dikes (cf. fig. 2.2). Dike <B> was not accessible for sampling.

2. Abbreviations for the basaltic rocks
   ON olivine nephelinite
   LI limburgite
   BA bassanite
   PT phonolitic tephrite
   nc not classified. Due to the intensive alteration the sample was not able to be classified according to the modern nomenclature for magmatic rocks.

3. and 4. The minerals listed are given according to estimated vol%. The components in parentheses are only of minor significance with respect to the total volume.
   The opaque minerals designated as 'ore' were not determined exactly with reflected-light microscopy. Several microprobe analyses of the ore grains yielded a titanomagnetite composition (dike <I> pyrite).

5. The particularities listed are described in more detail in the text.

6. Left of the virgule are the outcrop numbers, right is the number of studied thin sections in each case. The outcrops for the individual basalt dikes are arranged from north to south.

   ug subsurface (underground) exposure
   sf surface exposure

* Supplementary samples from KNIPPING (1984) and KNIPPING & HERRMANN (1985)

1	2	3	4	5	6
dike	type	phenocrysts	groundmass	remarks, particularities	outcrop/ thin sections
<A>	ON	olivine, augite	augite, nepheline, zeolites, ore, (apatite), (glass)	clusters of biotite; submicroscopic phases and glass, carbonates, and hypidiomorphic anhydrite in the groundmass	20-ug / 2
<C>	ON	olivine, augite	augite, nepheline, ore, (olivine), (apatite), (glass), (zeolite)	clusters of biotite, vesicles filled with carbonates and halite	21-ug / 2

Tab. 5.1 Petrography of the basalts, continued

1	2	3	4	5	6
dike	type	phenocrysts	groundmass	remarks, particularities	outcrop/ thin sections
\<D\>	BA	olivine, augite	augite, plagioklase, analcime, nepheline, biotite, ore, (glass), (apatite)	subsurface - no biotite; carbonates in cracks. Carbonates also with idiomorphic anhydrite partially replace the olivine and pyroxene phenocrysts. Surface - rhoenite as a product of the decomposition of biotite	15-ug / 4 45-sf / 3
\<E\>	ON	augite, olivine	augite, nepheline, ore, (glass),	clusters of biotite; microcrystalline carbonates and hypidiomorphic anhydrite at the contact in the matrix and in vesicles; xenoliths of peridotite. Outcrop 10-ug - augite phenocrysts in dark-brown, microlitic glass	10-ug / 2 22-ug / 2 1D-ug* / 10
\<F\>	BA	augite, olivine	augite, plagioklase, biotite	plagioclase as matrix; carbonates, anhydrite, (rhoenite) resulting from the decomposition of unidentifiable phenocrysts	2-ug / 2
\<G\>	ON	olivine, augite	augite, ore, (olivine), (nepheline), (glass), (apatite)	clusters of biotite with glass as matrix	4-ug / 2 51-ug / 2 18-ug / 2
\<H\>	BA	augite, olivine	plagioklase, nepheline, augite, ore, biotite, (analcime), (zeolite), (glass)	subsurface - clusters of biotite with plagioclase as matrix. Outcrop 52-ug - microlitic glass as groundmass. Surface - vesicles with zeolites; phyllosilicates in the groundmass	1-ug / 2 3-ug / 2 43-ug / 2 53-ug / 1 52-ug / 1
\<I\>	nc	(hornblende)	quartz, pyrite, glass, (plagioklase), (apatite)	highly weathered rock; carbonates, anhydrite, and halite in cracks and as vesicle filling. Rock cannot be classified based on thin sections	2D-ug* / 4 19-ug / 1
\<J\>	LI	(olivine), (augite)	augite, ore, glass, (plagioklase), (apatite)	the rare phenocrysts and groundmass minerals are only preserved as relicts; zeolites, anhydrite, carbonates, and mostly halite (with fluid inclusions) fill vesicles	57-ug / 1 36-ug / 1 17-ug / 2 16-ug / 2 34-ug / 1 35-ug / 1

1	2	3	4	5	6
dike	type	phenocrysts	groundmass	remarks, particularities	outcrop/ thin sections
<K>	LI	(olivine), (augite)	augite, ore, glass, (apatite)	even more weathered than <J>; relict biotite clusters; halite-filled amygdules permeate the rock	56-ug / 1 55-ug / 1 58-ug / 1 37-ug / 1
<L>	ON	olivine, augite	augite, nepheline, analcime, ore, (glass), (apatite)	relatively fresh rock; xenoliths of peridotite with ortho-pyroxenes, clinopyroxenes, and spinel	54-ug / 2 41-sf / 2 42-sf / 2
<M>	BA	olivine	augite, olivine, biotite, plagioklase, nepheline, ore, (apatite), (zeolite)	the olivine phenocrysts have been decomposed completely into chrysotile. Anhydrite and halite (in part with fluid and gas inclusions) fill the amygdules	5-ug / 2 6-ug / 2 7-ug / 2
<N>	BA	hornblende, augite, (olivine)	plagioklase, augite, biotite, nepheline, analcime, ore, (glass), (rhoenite), (apatite)	hornblendes and augites over 1.5 cm Ø; augite is very corroded; the rare olivines are essentially smaller and decomposed into chrysotile; in addition to nepheline and analcime, microlitic glass is found in the groundmass; anhydrite and halite in fine cracks. Surface - rhoenite from the decomposition of hornblende	13-ug / 4 32-ug / 1 44-sf / 2 8-ug / 2 27-ug / 1 24-ug / 2 23-ug / 1 30-ug / 1 29-ug / 1 28-ug / 1
<O>	PT	plagioklase	plagioklase, sanidine, nepheline, ore, (glass), (augite), (biotite), (apatite)	anhydrite, halite, and undetermined phyllosilicates and carbonates fill the amygdules and are disseminated in the matrix. Augite and hornblende only in outcrops 40 and 33; in outcrop 33-ug coarsely spathic plagioclase with sanadine and nepheline form the groundmass; sample 46/87 contains olivine (converted into 'iddingsite'). Strong variations in the vol% of the various minerals are observed from outcrop to outcrop	46-sf / 2 40-sf / 2 50-ug / 2 33-ug / 2 12-ug / 2 31-ug / 1
<P>	nc	olivine, augite	ore, glass, (plagioklase)	highly weathered rock; classification not possible using the microscopy.	46-sf / 2

brownish and, in contrast to the crystal cores, weakly pleochroitic. Greenish pyroxene cores are indicative of alkali dominance (aegirine-augite component). The crystals are generally about 1 to 3 mm in size. Yet, they may reach sizes of over 1 cm in the hornblende-bearing basanite of dike <N>. Corrosion of the rare pyroxenes in samples 40/21 and 33/723 (phonolitic tephrite, dike <O>) is usually very extensive. They appear to be replaced by hornblende.

When the groundmass of the basaltic rocks is not made up of microlitic glass, it is frequently composed of an acicular framework of hypidiomorphic to idiomorphic, brownish augite. The columnar crystals about 20 μm in size display the hour-glass structures characteristic of titanaugite. According to microprobe analyses it contains about 4 wt% TiO_2 (KNIPPING 1984).

In contrast to the augites the up to over 1.5 cm long amphiboles abundant in the basanite of dike <N> are unzoned and usually very fresh. The extinction angle (the angle between nγ and the crystallographic c-axis) was just a little over 0 °. Based on microprobe analyses the crystals contain between 4 and 5 wt% TiO_2. According to TRÖGER (1967) brown amphiboles with less than 5 wt% TiO_2 should not be referred to as kaersutite (e.g., VINX & JUNG 1977), but are products of the pargasite-hastingsite solid-solution series ('basaltic' hornblende). This mineral has an intense brown color and noteworthy pleochroism. The crystals are often surrounded by a fine dark rim of opaque phases.

The amphibole in the basanite of outcrop Lieshauk described by LAEMMLEN & MEISL (1975) was not observed (sample 45/94). A small number of smaller idiomorphic hornblende crystals (< 1 mm) are also found in the weathered rock of dike <I>. In dike <O> they appear to be the product of the decomposition of larger pyroxenes.

Feldspar
In the sample (31/68) of phonolitic tephrite from dike <O> elongate, polysynthetically twinned plagioclase crystals swim in a matrix of brown microlitic glass and very fine feldspars. The individual crystals are about 0.5 cm long. Their composition varies from andesine to labradorite as roughly determined with the angle of extinction (KÄDING 1962, p. 111). More rarely, tabular alkali feldspar, i.e. sanidine usually displaying Karlsbad twinning, is also observed in this rock. Sample 48/87 from Dreienberg near Friedewald has a very similar mineralogy and is thus ascribed to dike <O>.

Other samples of phonolitic tephrite have an aphyric, trachytic texture. In this case the rock consists of fine, flow-oriented plagioclase and sanidine crystals, nepheline, opaque minerals, and glass. The subsurface samples additionally contained halite- and carbonate-filled amygdules. In these samples sanidine can hardly be distinguished from plagioclase since it rarely occurs as larger individual plates.

In contrast, the phonolitic tephrite from outcrop 33-ug has an ophitic, intersertal texture. Chaotically oriented, inclusion-rich plagioclase together with sanidine and nepheline overgrow skeletal to idiomorphic hornblende crystals. The mesostasis consists of microcrystalline carbonates and phyllosilicates ('sericitization').

In the rocks characterized as olivine nephelinite plagioclase is lacking completely by definition. On the other hand, columnar plagioclase together with titanaugite forms the matrix of the basanites.

The vol% ratio of plagioclase to pyroxene can vary remarkably within one dike. For example, the groundmass of sample 53-ug from dike <H> contains little plagioclase and abundant augite crystals. In sample 43-sf from the same dike this relationship is exactly the opposite. Thus, this rock could be named alkali-olivine basalt.

Foids
Nepheline occurs rarely as idiomorphic crystals. It mostly acts as the filling between crystals of augite and plagioclase in the groundmass and is an inclusion-rich mineral phase which is very difficult to identify. In dikes with predominantly recrystallized, microlitic glass matrix it cannot be detected microscopically at all. Analcime, which is abundant in a similar way in some samples (e.g., dikes <D> and <L>), does not display its own crystal form and can only be distinguished from light-colored glass by its rarely occurring, very weak, abnormal birefringence.

Apatite
Apatite is also difficult to identify under the microscope in spite of its high relief and characteristically low interference colors of the 1st order. It usually occurs as acicular crystals less than 0.1 mm in size. Larger crystals only occur very rarely.

Biotite
Abundant biotite is found as a component of the groundmass primarily in the basanites. The characteristic mottling of the interference colors of the biotite is not always observed due to the intensive reddish-brown natural color and the smallness (about 0.2 mm in length) of the nonoriented crystals. Microcrystalline aggregates of reddish-brown, nearly opaque rhoenite occur rarely in surface samples resulting from the decomposition of biotite and hornblende.

Opaque minerals
Other components of the matrix are opaque minerals occurring in two generations. These components also referred to as 'ore' are usually titanomagnetite. In the highly weathered rock of dike <I> pyrite was also detected using a microprobe. A more exact determination of the opaque minerals with the reflected-light microscope was not done.

Glass
In unweathered dike rock, glass seldom occurs as fillings between crystals. However, reddish-brown, transparent glass does form a 1- to 2-cm-wide margin in the unweathered basaltic rocks at the contact to the salt rock.

The phenocrysts contained in this glass are only slightly corroded. Recrystallized, microlitic glass can also be rock-forming, primarily in the limburgites and the unclassified silicate rocks. This glass is always opaque and brown in color due to its *submicroscopic components.*

Amygdules, Ocelli

Zeolites, which were not determined more specifically, were observed as amygdule fillings in a few surface and subsurface samples. In the limburgites and the undifferentiated samples the filling in the abundant amygdules was halite. The same observation was made in the fresh samples near the contact to the evaporites. In many cases the sharp contact between the glassy groundmass of the silicate rock and the amygdule filling was formed by fine-grained to idiomorphic anhydrite and/or carbonates and phyllosilicates (fig. 5.1). This crystallization sequence is also found in the abundant cracks and fractures in the intrusive rocks and at the contact basalt - salt rock.

In contrast to the amygdules or vesicles, clusters of aggregates of various acicular silicate minerals are found in unweathered rock, primarily in olivine nephelinite. These round, irregularly shaped clusters or 'ocelli' are distinguished by a smooth transition from the silicate rock matrix to the halite filling (WIMMENAUER 1985).

The ocelli consist of acicular biotite and augite (ON) or plagioclase (BA). The crystals are commonly oriented tangentially to the cluster filling. Nepheline, glass, and microcrystalline phyllosilicates as well as granular, dendritic, or round, radiating ores form the matrix. Like the amygdules, the margins of the clusters are lined with glass, carbonates, and/or anhydrite. The filling consists of halite (fig. 5.2). Similar occurrences in nephelinites at the Buggingen salt mine are discussed by HURRLE (1976).

Gases

The nearly submicroscopic, very high relief inclusions in the halite fillings of the amygdules and clusters are possibly gas inclusions. The cleavage typical for halite usually appears only as shadows.

Sulfur

Since native sulfur was not observed in any of the basalt thin sections, it must be disseminated through the silicate rock in submicroscopic phases. Thus, the macroscopic sulfur crystals in the basalt (e.g., outcrop 8-ug) are only local accumulations.

Evidence of the presence of further basalt occurrences in the Werra mining district was provided by extensive ΔT measurements at the surface with a proton magnetometer (SIEMENS 1971). The results show that dike <O> (normal magnetization) possibly extends to the quarry near Friedewald (outcrop 46-sf). Further evidence of such is the similarity between samples 46/87 and 31/68 regarding texture and composition.

Nevertheless, sample 46/86 (inverse magnetization) was given a new designation (dike <P>) since there were no mineralogical criteria for associating it with the rock of sample 46/87. KÄDING (1962) described three NS-striking basalt dikes in the quarry near Friedewald, which are no longer visible due to poor exposure. While sample 46/87 is from one of these dikes (see KÄDING 1962, fig. 21), sample 46/86 obviously belongs to the surrounding basalt.

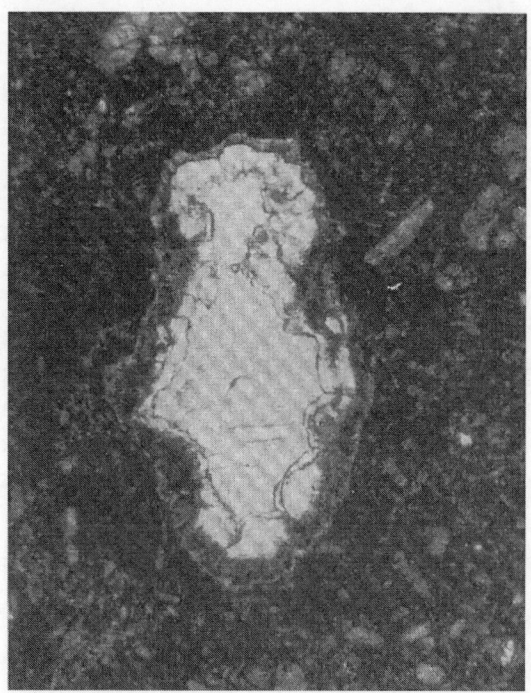

0.25 mm

Fig. 5.1 Olivine nephelinite from outcrop 20-ug, dike <A>. Idiomorphic anhydrite crystals have grown into an amygdule together with xenomorphic carbonates. The amygdule is filled with halite. The impurities in the halite were produced during the making of the thin section.
(nonpolarized)

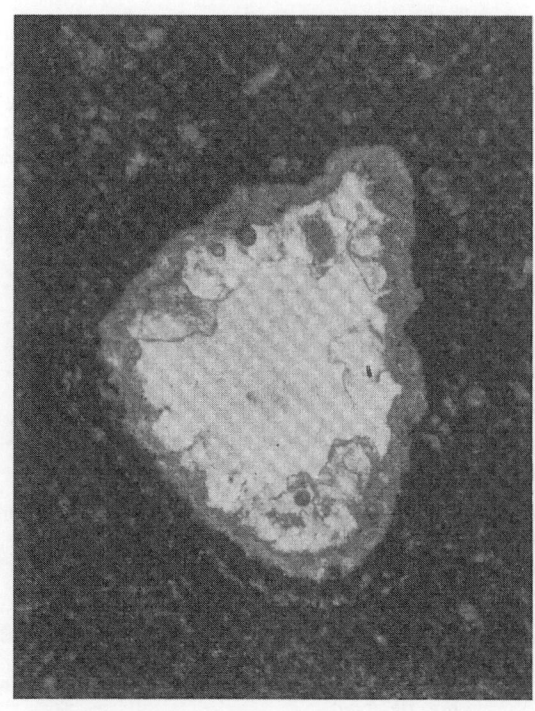

0.25 mm

Fig. 5.2 Same as fig. 5.1. Ocellus filled with halite. The margins of the ocellus are composed of glass and xenomorphic carbonates. Tangentially arranged biotites and phyllosilicates form the transition from the basalt matrix to the glassy margin of the ocellus (section is a little too thick).
(nonpolarized)

5.2 Chemistry

The major, minor, and trace elements of the chloride-free basalt samples are given in table 5.2. The geographical locations of the outcrops in the study area can be seen in fig. 2.2.

The amounts (mass fraction in %, wt%) of the components ΣH_2O, ΣC as CO_2, and S in the subsurface samples are striking and unusually high for basalts. As microscopy has shown, several samples contain analcime or amphiboles and biotite as well as frequently undeterminable phyllosilicates and considerable amounts of glass. These components may be responsible for the high amounts of H_2O. Carbon is fixed in the carbonate minerals. Sulfur usually occurs as sulfate (anhydrite), but also as native sulfur and sulfide. Due to their fine dissemination in the groundmass of the basalt these mineral components could not be completely removed before the chemical components had been determined.

The sample-by-sample normalization of the analyses for water-, carbonate-, and sulfur-free amounts did not produce any results comparable with the respective surface samples. Causes for this are the mobilization of some components by fluid phases, the variable composition of secondary phyllosilicates and carbonates, and the changing ratio of sulfates and sulfides to native sulfur (chapter 5.3; KNIPPING 1984).

Table 5.2   Major, minor, and trace components in the basaltic rocks

Explanations:

The samples within the individual basalt dikes are arranged from north to south. Since the amounts of sulfur vary between about 50 and 22 100 µg S/g rock, they are given under minor or trace elements. In several subsurface samples sulfur occurs in the form of sulfate and possibly sulfide. The totals from the analyses which are in part too low can be explained by the fact that the total amounts of sulfur are given as native sulfur (S), i.e. without the oxygen of SO_4 constituents.

<A> to <P>   numeration of dikes (cf. fig. 2.2). Dike <B> was not accessible for sampling.

Abbreviations for the basaltic rocks

ON   olivine nephelinite
LI   limburgite
BA   basanite
PT   phonolitic tephrite
nc   not classified. Due to the intensive alteration the sample was not able to be classified according to the modern nomenclature for magmatic rocks.

* Supplementary samples from KNIPPING (1984) and KNIPPING & HERRMANN (1985)

ug   subsurface (underground) exposure
sf   surface exposure

The values for the subsurface basalts are from chloride-free sample material.

**   Data under major and minor or trace components
n.d.   not detected

dike	<A>	<C>	<D>			<E>
rock type	ON	ON	BA			ON
sample	20/43 ug	21/46 ug	15/33a ug	15/33b ug	45/94 sf	10/22 ug
major and minor components, mass fraction in % (wt%)						
SiO_2	39.4	40.1	40.1	40.2	41.6	37.9
TiO_2	3.0	2.8	3.0	3.1	2.8	2.5
Al_2O_3	11.0	11.2	11.9	11.7	10.7	12.1
Fe_2O_3	5.5	4.4	3.9	4.1	5.0	4.9
FeO	6.7	6.5	6.6	6.3	6.2	6.3
MnO	0.1_7	0.1_9	0.1_7	0.1_6	0.1_8	0.1_9
MgO	12.6	12.1	11.0	10.3	10.4	13.4
CaO	10.8	12.1	11.2	10.8	12.4	9.4
Na_2O	2.0	2.6	2.1	2.2	3.2	2.3
K_2O	1.6	2.1	2.1	2.1	1.4	1.9
ΣH_2O	5.1	3.1	2.9	3.5	2.0	4.2
P_2O_5	0.5_7	0.6_7	0.5_0	0.5_5	0.4_1	1.1
ΣC as CO_2	1.2	1.5	3.8	3.5	3.3	3.2
S	**	0.12	0.16	0.52	**	0.21
total	99.6	99.5	99.4	99.0	99.6	99.6
ΣFe as Fe_2O_3	12.9	11.6	11.2	11.1	11.9	11.9
trace components, μg element/g rock (ppm)						
S	760	**	**	**	870	**
Sc	25.3	26.6	29.9	27.1	31.0	26.0
Cr	410	270	410	350	450	390
Co	58.4	53.0	55.2	54.0	58.0	54.0
Ni	281	267	211	200	232	241
La	53	72	46	46	43	73
Ce	100	116	91	75	89	126
Sm	9.2	9.9	8.2	8.6	7.7	10.0
Eu	2.5	3.0	2.4	2.6	2.3	3.0
Tb	1.0	1.3	0.9	1.1	0.9	1.3
Yb	1.6	2.1	1.9	1.5	1.7	2.2
Lu	0.25	0.32	0.26	0.23	0.26	0.34
Hf	5	5	5	4	5	6

Table 5.2 Major, minor, and trace components in the basaltic rocks, continued

dike	<E>		<F>	<G>		
rock type	ON		BA	ON		
sample	22/48 ug	1D/12* ug	2/5 ug	4/11 ug	51/89 ug	18/40 ug
major and minor components, mass fraction in % (wt%)						
SiO_2	37.7	41.1	40.9	41.3	41.1	37.2
TiO_2	2.7	2.4	3.1	3.3	3.2	3.3
Al_2O_3	11.4	11.5	12.9	11.9	12.0	11.6
Fe_2O_3	4.2	4.2	5.0	5.8	5.1	5.6
FeO	6.1	7.0	5.0	6.1	6.7	6.3
MnO	0.1_8	0.2_0	0.1_5	0.1_8	0.1_8	0.1_7
MgO	10.8	12.3	8.8	9.8	9.4	8.7
CaO	10.8	11.9	10.3	10.1	11.6	9.9
Na_2O	3.2	2.9	2.0	1.6	1.8	1.0
K_2O	2.0	1.6	3.0	2.7	2.6	2.6
ΣH_2O	4.3	3.7	3.7	5.2	3.8	4.9
P_2O_5	1.0	0.6_5	0.7_5	0.5_6	0.6_7	0.6_1
ΣC as CO_2	4.7	0.2	3.9	0.91	0.68	6.2
S	**	**	**	0.28	0.12	0.97
total	99.1	99.6	99.5	99.7	99.0	99.1
Fe as Fe_2O_3	11.0	12.0	10.6	12.6	12.5	12.6
trace components, μg element/g rock (ppm)						
S	950	910	770	**	**	**
Sc	24.8	26.7	26.6	30.4	31.9	27.3
Cr	400	470	220	260	250	220
Co	48.7	55.0	42.4	50.1	51.5	46.0
Ni	209	243	123	125	112	89
La	79	69	68	51	47	51
Ce	145	105	141	115	96	100
Sm	10.3	12.2	10.3	9.6	8.1	9.1
Eu	3.0	3.0	3.1	2.8	2.7	2.6
Tb	1.1	1.6	1.1	1.2	0.7	1.0
Yb	2.3	2.4	2.3	2.1	1.8	2.1
Lu	0.26	0.29	0.34	0.31	0.26	0.31
Hf	6	7	7	7	5	6

dike	<H>					<I>
rock type	BA					nc
sample	1/2 ug	3/7 ug	43/92 sf	53/96 ug	52/95 ug	2D/24* ug
major and minor components, mass fraction in % (wt%)						
SiO_2	41.5	39.9	43.9	42.7	42.0	39.8
TiO_2	3.2	3.3	3.1	3.3	3.1	2.7
Al_2O_3	12.4	12.9	13.3	12.5	12.5	12.8
Fe_2O_3	5.4	5.5	4.1	4.8	5.2	5.4
FeO	6.4	6.5	6.1	4.9	6.3	5.2
MnO	0.1_7	0.1_8	0.2_1	0.1_3	0.1_8	0.1_8
MgO	9.9	10.0	7.0	11.2	9.1	6.5
CaO	11.4	11.6	11.1	9.2	11.2	6.5
Na_2O	1.8	1.9	3.5	1.6	2.2	1.8
K_2O	2.5	2.0	2.4	2.7	2.0	3.8
ΣH_2O	3.5	3.7	2.2	4.2	4.2	5.0
P_2O_5	0.6_8	0.6_5	0.5_6	0.5_7	0.6_4	0.6_2
ΣC as CO_2	1.0	1.6	2.1	1.1	0.41	9.5
S	**	0.10	**	0.17	0.22	0.20
total	99.9	99.8	99.6	99.1	99.3	100.0
Σ Fe as Fe_2O_3	12.5	12.7	10.9	10.2	12.2	11.2
trace components, μg element/g rock (ppm)						
S	970	**	170	**	**	**
Sc	32.7	33.5	26.9	31.2	30.0	12.8
Cr	320	300	230	270	260	n.d.
Co	52.6	53.0	42.5	42.1	48.8	26.6
Ni	139	132	107	102	108	<10
La	. 52	49	63	43	49	75
Ce	96	94	128	89	100	117
Sm	9.9	9.2	9.4	7.4	8.3	15.7
Eu	3.0	2.9	2.9	2.4	2.8	3.8
Tb	1.7	1.1	1.0	0.7	0.8	2.2
Yb	2.4	2.0	2.2	1.8	1.9	2.9
Lu	0.27	0.32	0.33	0.23	0.25	0.26
Hf	6	6	7	5	4	10

Table 5.2 Major, minor, and trace components in the basaltic rocks, continued

dike	\<I\>	\<J\>				
rock type	nc	LI				
sample	19/41 ug	57/140 ug	36/78 ug	17/39 ug	16/37 ug	34/75 ug
major and minor components, mass fraction in % (wt%)						
SiO_2	47.8	41.7	42.6	40.4	41.1	41.7
TiO_2	0.9_7	3.4	3.7	3.4	3.5	3.3
Al_2O_3	12.0	12.8	14.6	12.5	12.4	12.7
Fe_2O_3	1.3	5.4	5.0	6.3	6.0	5.8
FeO	4.9	7.0	5.7	5.8	6.4	5.7
MnO	0.1_3	0.2_0	0.1_7	0.1_8	0.1_8	0.1_6
MgO	9.4	8.0	5.8	8.7	8.7	10.2
CaO	0.8_2	10.9	8.6	9.2	10.7	8.8
Na_2O	0.3_0	0.8_2	2.8	2.1	1.9	1.4
K_2O	2.6	3.6	2.9	2.8	2.8	3.4
ΣH_2O	4.7	3.9	5.7	5.9	4.6	5.0
P_2O_5	0.2_8	0.7_6	0.8_3	0.7_9	0.7_5	0.5_9
ΣC as CO_2	13.6	0.5_3	0.5_8	0.9_2	0.4_5	0.7_5
S	0.38	0.10	0.20	0.24	**	0.22
total	99.2	99.1	99.2	99.2	99.5	99.7
ΣFe as Fe_2O_3	6.7	13.2	11.3	12.7	13.1	12.1
trace components, μg element/g rock (ppm)						
S	**	**	**	**	900	**
Sc	11.3	25.9	19.0	22.5	25.2	28.0
Cr	60	120	30	n.d.	150	250
Co	16.7	47.6	33.6	43.3	45.4	48.8
Ni	32	74	32	51	74	132
La	38	62	64	70	65	62
Ce	69	126	121	111	130	124
Sm	6.8	9.7	9.8	11.8	10.8	10.2
Eu	1.5	3.2	3.1	3.4	3.2	3.0
Tb	0.9	0.9	1.0	1.5	1.1	1.1
Yb	2.3	2.2	2.1	2.3	2.5	2.2
Lu	0.41	0.30	0.31	0.36	0.34	0.30
Hf	4	5	6	6	7	7

dike	<J>	<K>				<L>
rock type	LI	LI				ON
sample	35/76 ug	56/99 ug	55/98 ug	58/141 ug	37/79 ug	54/97 ug
major and minor components, mass fraction in % (wt%)						
SiO_2	40.3	43.2	44.8	43.0	40.1	42.5
TiO_2	3.5	3.5	3.3	3.5	3.4	3.5
Al_2O_3	12.7	14.8	15.2	14.6	13.0	13.8
Fe_2O_3	6.8	6.1	5.8	4.9	6.2	4.6
FeO	6.6	5.2	5.2	5.9	6.3	6.3
MnO	0.2_1	0.2_1	0.1_7	0.1_8	0.1_9	0.1_8
MgO	8.3	6.0	4.5	6.8	8.4	8.2
CaO	11.0	7.9	9.2	7.5	9.9	10.0
Na_2O	1.8	2.2	4.0	1.5	2.1	3.4
K_2O	2.9	3.8	2.3	3.5	3.0	2.7
ΣH_2O	3.6	3.7	2.3	4.9	4.9	2.7
P_2O_5	0.8_2	0.7_7	0.9_0	0.7_7	0.7_9	0.6_6
ΣC as CO_2	0.7_3	1.2	1.2	1.7	0.9_1	0.2_6
S	0.11	0.32	0.14	0.17	0.17	0.12
Summe	99.4	98.9	99.0	98.9	99.4	98.9
ΣFe as Fe_2O_3	14.1	11.9	11.6	11.5	13.2	11.6
trace components, μg element/g rock (ppm)						
S	**	**	**	**	**	**
Sc	24.9	18.9	17.9	19.1	22.7	28.3
Cr	30	20	20	20	20	200
Co	47.2	33.5	31.8	32.8	44.4	45.1
Ni	59	22	19	26	65	115
La	70	61	69	61	69	55
Ce	114	118	135	121	109	111
Sm	11.7	9.2	10.0	9.2	11.3	9.0
Eu	3.3	3.1	3.2	3.1	3.2	2.8
Tb	1.7	0.8	0.8	0.8	1.5	1.0
Yb	2.2	2.1	2.4	2.1	2.4	2.2
Lu	0.27	0.27	0.32	0.29	0.28	0.36
Hf	10	5	5	5	10	6

Table 5.2 Major, minor, and trace components in the basaltic rocks, continued

dike	<L>		<M>			<N>
rock type	ON		BA			BA
sample	41/90 sf	42/91 sf	5/13 ug	6/14 ug	7/15 ug	13/30 ug
major and minor components, mass fraction in % (wt%)						
SiO_2	41.3	41.8	42.1	42.2	42.4	39.4
TiO_2	2.9	2.6	2.8	2.8	2.8	3.8
Al_2O_3	11.9	10.6	12.6	12.1	12.1	12.5
Fe_2O_3	4.4	4.1	3.8	3.9	3.7	6.5
FeO	7.0	7.5	6.4	6.7	6.5	5.2
MnO	0.2_0	0.2_1	0.1_5	0.1_4	0.1_4	0.1_6
MgO	11.7	13.1	11.3	12.3	12.1	10.5
CaO	12.3	12.1	9.7	9.4	9.6	10.5
Na_2O	3.6	3.6	1.9	1.9	1.9	1.7
K_2O	1.4	1.2	1.9	2.3	1.9	3.2
ΣH_2O	1.9	2.0	5.2	3.9	4.7	3.1
P_2O_5	1.1	1.0	0.3_9	0.3_5	0.3_7	0.5_3
ΣC as CO_2	0.2_0	0.1_1	0.8_5	1.3	0.85	2.4
S	**	**	0.22	0.12	0.16	0.24
total	99.9	99.9	99.3	99.4	99.2	99.7
ΣFe as Fe_2O_3	12.2	12.4	10.9	11.3	10.9	12.3
trace components, μg element/g rock (ppm)						
S	160	120	**	**	**	**
Sc	28.0	25.8	27.5	27.1	26.6	36.6
Cr	460	610	400	370	410	190
Co	53.0	57.1	52.9	55.0	52.9	51.9
Ni	233	339	210	239	279	86
La	78	88	39	38	36	46
Ce	154	162	71	78	70	85
Sm	11.0	11.5	7.4	7.0	6.9	10.0
Eu	3.3	3.2	2.3	2.1	2.2	2.7
Tb	1.1	1.5	1.3	0.8	1.2	1.0
Yb	2.2	2.5	1.7	1.6	1.6	1.9
Lu	0.31	0.30	0.24	0.25	0.28	0.30
Hf	6	8	5	4	5	5

dike	<N>					
rock type	BA					
sample	13/27 ug	32/70 ug	44/93 sf	8/17 ug	27/59 ug	24/55 ug
major and minor components, mass fraction in % (wt%)						
SiO_2	39.9	41.7	42.4	41.6	41.4	41.2
TiO_2	3.7	3.4	3.6	3.6	3.7	3.7
Al_2O_3	13.4	13.6	13.6	12.9	13.9	12.9
Fe_2O_3	4.9	5.8	5.0	5.9	6.1	6.2
FeO	6.2	6.5	7.1	6.4	6.2	5.2
MnO	0.1_6	0.2_1	0.2_0	0.1_9	0.2_0	0.1_3
MgO	11.2	7.4	7.8	9.0	8.1	10.1
CaO	11.0	11.1	11.7	10.5	11.0	8.8
Na_2O	1.8	2.2	3.5	1.7	3.4	2.3
K_2O	3.0	2.7	1.6	3.6	1.7	2.8
ΣH_2O	2.8	3.1	2.5	2.7	2.3	3.4
P_2O_5	0.5_7	0.8_2	0.8_0	0.8_3	0.6_1	0.6_5
ΣC as CO_2	0.7_3	0.6_1	0.1_0	0.2_0	0.5_5	0.3_7
S	0.14	**	0.18	0.12	0.83	
total	99.5	99.3	99.9	99.3	99.3	98.6
ΣFe as Fe_2O_3	11.8	13.0	12.9	13.0	13.0	12.0
trace components, μg element/g rock (ppm)						
S	**	**	270	**	**	**
Sc	36.0	24.1	27.6	29.3	29.5	28.4
Cr	210	110	100	150	130	140
Co	51.6	46.8	46.6	49.5	48.4	45.8
Ni	80	68	78	83	80	70
La	42	63	63	57	58	52
Ce	81	128	104	122	100	106
Sm	9.0	10.5	11.6	10.6	11.1	9.8
Eu	2.7	3.2	3.3	3.2	3.2	2.8
Tb	1.0	1.0	1.5	1.2	1.7	1.1
Yb	1.5	2.2	2.1	2.3	2.1	2.0
Lu	0.27	0.29	0.24	0.35	0.26	0.29
Hf	4	6	8	7	9	7

Table 5.2 Major, minor, and trace components in the basaltic rocks, continued

dike	<N>				<O>	
rock type	BA				PT	
sample	23/52 ug	30/65 ug	29/63 ug	28/61 ug	46/87 sf	40/21 sf
major and minor components, mass fraction in % (wt%)						
SiO_2	40.4	41.5	40.8	40.1	51.9	45.9
TiO_2	3.6	3.8	3.8	3.7	2.5	3.0
Al_2O_3	12.6	13.0	13.8	12.9	16.2	16.3
Fe_2O_3	7.2	5.7	6.2	7.0	7.6	4.9
FeO	5.0	6.1	6.2	5.8	2.5	5.5
MnO	0.2_0	0.1_8	0.1_7	0.1_9	0.1_1	0.2_3
MgO	9.2	9.7	9.9	10.5	2.4	4.0
CaO	9.4	10.2	9.6	10.2	6.8	9.1
Na_2O	1.6	2.0	1.7	1.8	4.1	4.7
K_2O	3.4	2.4	2.4	2.7	0.8_4	2.3
ΣH_2O	2.9	3.8	4.5	3.4	4.3	1.9
P_2O_5	0.6_9	0.6_5	0.6_2	0.6_4	0.4_8	1.1
ΣC as CO_2	0.3_3	0.5_4	0.2_3	0.3_6	0.08	0.2_9
S	2.2	0.18	0.14	0.28	**	**
total	98.7	99.8	100.1	99.6	99.8	99.2
ΣFe as Fe_2O_3	12.8	12.5	13.1	13.4	10.4	11.0
trace components, μg element/g rock (ppm)						
S	**	**	**	**	130	160
Sc	28.6	32.2	33.1	33.3	14.8	13.0
Cr	120	150	130	190	40	10
Co	49.9	51.4	50.7	51.5	31.5	25.0
Ni	75	83	75	90	94	15
La	59	53	50	51	31	84
Ce	100	110	93	104	57	160
Sm	11.0	10.4	9.2	10.0	6.6	15.5
Eu	3.1	3.1	2.9	2.9	2.6	4.6
Tb	1.6	1.2	1.1	1.3	0.8	1.6
Yb	2.1	2.1	1.9	2.0	1.8	3.1
Lu	0.28	0.30	0.26	0.29	0.22	0.39
Hf	8	6	6	6	3	9

dike	<O>				<P>
rock type	PT				nc
sample	50/85 ug	33/723 ug	12/25 ug	31/68 ug	46/86 sf
major and minor components, mass fraction in % (wt%)					
SiO_2	55.2	43.4	52.8	53.7	47.1
TiO_2	1.2	2.8	1.2	1.3	2.5
Al_2O_3	18.5	15.5	17.9	17.9	12.7
Fe_2O_3	3.0	5.6	3.0	1.3	5.6
FeO	2.1	4.6	2.6	3.1	5.4
MnO	0.1_3	0.2_5	0.2_3	0.2_1	0.1_7
MgO	3.6	4.3	2.3	3.3	7.3
CaO	2.8	8.6	3.4	2.0	10.4
Na_2O	6.1	3.7	4.4	3.3	2.8
K_2O	3.6	3.3	5.0	3.8	1.4
ΣH_2O	1.9	2.5	2.4	4.0	3.1
P_2O_5	0.5_1	1.1	0.4_7	0.3_6	0.5_5
ΣC as CO_2	0.1_3	3.6	3.4	4.2	0.06
S	0.12	0.12	**	0.29	**
total	98.9	99.4	99.1	98.8	99.1
ΣFe as Fe_2O_3	5.3	10.7	5.9	4.7	11.6
trace components, μg element/g rock (ppm)					
S	**	**	410	**	50
Sc	3.2	14.3	3.0	3.0	19.1
Cr	10	20	n.d.	n.d.	210
Co	3.1	23.5	1.8	1.9	45.0
Ni	30	15	11	n.d.	173
La	111	86	128	122	41
Ce	243	178	202	196	52
Sm	17.7	15.9	22.4	19.4	6.6
Eu	5.7	4.6	6.5	5.6	1.8
Tb	1.5	1.6	2.8	2.3	0.7
Yb	4.1	3.2	4.2	4.0	1.8
Lu	0.49	0.44	0.41	0.44	0.23
Hf	9	10	18	15	3

The mean values of the chemical composition of the dikes are compiled in table 5.3. At least three samples were analyzed in each case. Due to the small number of samples for individual dikes it was not practical to calculate the standard deviation of the chemical components. Great deviations in the mean values of the composition of the differing dikes can occasionally be recognized, particularly for water, carbon (fixed in the rock as Mg-Ca-Fe carbonates), sulfur (mostly as SO_4 in anhydrite), and some trace elements (e.g., chrome and nickel). These data illustrate the range of variation in chemical composition of the basalts possible within one dike or rock type. They consequently confirm the observations made with the microscope.

Calculating the composition
In contrast to fresh basalts and surface samples, many subsurface samples have a Na/K ratio of < 1. Hence, the (Na+K)/Si ratio (RINGWOOD 1975) was not able to be used for supplementing the petrographic classification of the subsurface basalts based on microscopy. The calculation of the CIPW norm is unsuitable as well because slight deviations in the amounts of alkali elements give rise to great differences in the amounts of CIPW-normative minerals. Consequently, CIPW norm was only calculated for the surface samples having low amounts of H_2O, C, and S (table 5.4). The results agree well with the data of WEDEPOHL (1983) for basanites and olivine nephelinites west of the Werra mining district (table 5.5). Basanites from dikes <D> and <H> (table 5.4) show a clear tendency toward alkali-olivine basalts. As anticipated, the phonolitic tephrite is not comparable with the true basaltic rocks.

The CIPW norm considers mainly the leucocratic minerals in a rock. Yet, it does appear to be necessary to consider above all the mafic minerals, especially in the case of the hornblende-bearing basanites from dike <O>. Therefore, the chemical compo-

---

Table 5.3   Mean values and range of variation of the chemical composition of the basalt dikes

Explanations:

<D>   numeration of dikes (cf. fig. 2.2). The mean values of the chemical composition were calculated with the values given in table 5.2 for those dikes from which more than two outcrops were sampled.

Abbreviations for the basaltic rocks

ON   olivine nephelinite
LI   limburgite
BA   basanite
PT   phonolitic tephrite
nc   not classified. Due to the intensive alteration the sample was not able to be classified according to the modern nomenclature for magmatic rocks.

* Supplementary samples from KNIPPING (1984) and KNIPPING & HERRMANN (1985)

The values are from chloride-free sample material.

dike	<D>			<E>		
rock type/ number of samples	BA/3			ON/3		

major and minor components, mass fraction in % (wt%).

	$\bar{x}$	x_{min}	-	x_{max}	$\bar{x}$	x_{min}	-	x_{max}
SiO_2	40.6	40.1	-	41.6	38.9	37.7	-	41.1
TiO_2	3.0	2.8	-	3.1	2.5	2.4	-	2.7
Al_2O_3	11.4	10.7	-	11.9	11.7	11.4	-	12.1
Fe_2O_3	4.3	3.9	-	5.0	4.4	4.2	-	4.9
FeO	6.4	6.2	-	6.6	6.5	6.1	-	7.0
MnO	0.17	0.16	-	0.18	0.19	0.18	-	0.20
MgO	10.6	10.3	-	11.0	12.2	10.8	-	13.4
CaO	11.5	10.8	-	12.4	10.7	9.41	-	11.9
Na_2O	2.5	2.1	-	3.2	2.8	2.3	-	3.2
K_2O	1.9	1.4	-	2.1	1.8	1.6	-	2.0
ΣH_2O	2.8	2.0	-	3.5	4.1	3.7	-	4.3
P_2O_5	0.49	0.41	-	0.55	0.92	0.65	-	1.1
ΣC as CO_2	3.5	3.3	-	3.8	2.7	0.19	-	4.7
S [$\mu g/g$]	2580	870	-	5240	1330	910	-	2130

trace components, μg element/g rock (ppm)

Sc	29.3	27.1	-	31.0	25.8	24.8	-	26.7
Cr	400	350	-	450	420	390	-	470
Co	55.7	54.0	-	58.0	52.6	48.7	-	55.0
Ni	214	200	-	232	231	209	-	243
La	45	43	-	46	74	69	-	79
Ce	85	75	-	91	125	105	-	145
Sm	8.2	7.7	-	8.6	10.8	10.0	-	12.2
Eu	2.4	2.3	-	2.6	3.0	3.0	-	3.0
Tb	0.8	0.6	-	0.9	1.3	1.1	-	1.6
Yb	1.7	1.5	-	1.9	2.3	2.2	-	2.4
Lu	0.25	0.23	-	0.26	0.30	0.26	-	0.34
Hf	5	4	-	5	6	6	-	7

Table 5.3 Mean values of the chemical composition, continued

dike	<G>			<H>		
rock type/ number of samples	ON/3			BA/5		

major and minor components, mass fraction in % (wt%)

	$\bar{x}$	x_{min}	- x_{max}	$\bar{x}$	x_{min}	- x_{max}
SiO_2	39.9	37.2	- 41.3	42.0	39.9	- 43.9
TiO_2	3.3	3.2	- 3.3	3.2	3.1	- 3.3
Al_2O_3	11.8	11.6	- 12.0	12.7	12.4	- 13.3
Fe_2O_3	5.5	5.1	- 5.8	5.0	4.1	- 5.5
FeO	6.4	6.1	- 6.7	6.0	4.9	- 6.5
MnO	0.18	0.17	- 0.18	0.17	0.13	- 0.21
MgO	9.3	8.7	- 9.8	9.4	7.0	- 11.2
CaO	10.5	9.9	- 11.6	10.9	9.2	- 11.6
Na_2O	1.5	1.0	- 1.8	2.2	1.6	- 3.5
K_2O	2.6	2.6	- 2.7	2.3	2.0	- 2.7
ΣH_2O	4.6	3.8	- 5.2	3.6	2.2	- 4.2
P_2O_5	0.61	0.56	- 0.67	0.62	0.56	- 0.68
ΣC as CO_2	2.6	0.68	- 6.2	1.2	0.41	- 2.1
S [$\mu g/g$]	4553	1220	- 9650	1210	170	- 2220

trace components, μg element/g rock (ppm)

Sc	29.9	27.3	- 31.9	30.8	26.9	- 33.5
Cr	240	220	- 260	280	230	- 320
Co	49.2	46.0	- 51.5	47.8	42.1	- 53.0
Ni	109	89	- 125	118	102	- 139
La	49	47	- 51	51	43	- 63
Ce	104	96	- 115	102	89	- 128
Sm	8.9	8.1	- 9.6	8.8	7.4	- 9.9
Eu	2.7	2.6	- 2.8	2.8	2.4	- 3.0
Tb	1.0	0.7	- 1.2	1.1	0.7	- 1.7
Yb	2.0	1.8	- 2.1	2.1	1.8	- 2.4
Lu	0.29	0.26	- 0.31	0.28	0.23	- 0.33
Hf	6	5	- 7	6	4	- 7

dike	<J>			<K>		
rock type/ number of samples	LI/6			LI/4		

major and minor components, mass fraction in % (wt%)

	$\bar{x}$	x_{min}	-	x_{max}	$\bar{x}$	x_{min}	-	x_{max}
SiO_2	41.3	40.3	-	42.6	42.8	40.1	-	44.8
TiO_2	3.5	3.3	-	3.7	3.4	3.3	-	3.5
Al_2O_3	13.0	12.4	-	14.6	14.4	13.0	-	15.2
Fe_2O_3	5.9	5.0	-	6.8	5.8	4.9	-	6.2
FeO	6.2	5.7	-	7.0	5.7	5.2	-	6.3
MnO	0.18	0.16	-	0.21	0.19	0.17	-	0.21
MgO	8.3	5.8	-	10.2	6.4	4.5	-	8.4
CaO	9.9	8.6	-	11.0	8.6	7.5	-	9.9
Na_2O	1.8	0.8	-	2.8	2.5	1.5	-	4.0
K_2O	3.1	2.8	-	3.6	3.2	2.3	-	3.8
ΣH_2O	4.8	3.6	-	5.9	4.0	2.3	-	4.9
P_2O_5	0.76	0.59	-	0.83	0.81	0.77	-	0.90
ΣC as CO_2	0.66	0.45	-	0.92	1.3	0.91	-	1.7
S [$\mu g/g$]	1595	900	-	2390	1965	1360	-	3150

trace components, μg element/g rock (ppm)

Sc	24.2	19,0	-	28.0	19.6	17.9	-	22.7
Cr	100	n.d.	-	250	20	20	-	20
Co	44.3	33.6	-	48.8	35.6	31.8	-	44.4
Ni	70	32	-	132	33	19	-	65
La	65	62	-	70	65	61	-	69
Ce	121	111	-	130	120	109	-	135
Sm	10.7	9.7	-	11.8	9.9	9.2	-	11.3
Eu	3.2	3.0	-	3.4	3.2	3.1	-	3.2
Tb	1.2	0.9	-	1.7	1.0	0.8	-	1.5
Yb	2.2	2.1	-	2.5	2.2	2.1	-	2.4
Lu	0.31	0.27	-	0.36	0.29	0.27	-	0.32
Hf	7	5	-	10	6	5	-	10

Table 5.3 Mean values of the chemical composition, continued

dike	<L>			<M>		
rock type/ number of samples	ON/3			BA/3		

major and minor components, mass fraction in % (wt%)

	$\overline{x}$	x_{min}	- x_{max}	$\overline{x}$	x_{min}	- x_{max}
SiO_2	41.9	41.3	- 42.5	42.2	42.1	- 42.4
TiO_2	3.0	2.6	- 3.5	2.8	2.8	- 2.8
Al_2O_3	12.1	10.6	- 13.8	12.3	12.1	- 12.6
Fe_2O_3	4.4	4.1	- 4.6	3.8	3.7	- 3.9
FeO	6.9	6.3	- 7.5	6.5	6.4	- 6.7
MnO	0.20	0.18	- 0.21	0.14	0.14	- 0.15
MgO	11.0	8.2	- 13.1	11.9	11.3	- 12.3
CaO	11.5	10.0	- 12.3	9.6	9.4	- 9.7
Na_2O	3.5	3.4	- 3.6	1.9	1.9	- 1.9
K_2O	1.8	1.2	- 2.7	2.0	1.9	- 2.3
ΣH_2O	2.2	1.9	- 2.7	4.6	3.9	- 5.2
P_2O_5	0.92	0.66	- 1.1	0.37	0.35	- 0.39
ΣC as CO_2	0.19	0.11	- 0.26	1.0	0.85	- 1.3
S [$\mu g/g$]	507	120	- 1240	1657	1200	- 2180

trace components, μg element/g rock (ppm)

Sc	27.4	25.8	- 28.3	27.1	26.6	- 27.5
Cr	420	200	- 610	390	370	- 410
Co	51.7	45.1	- 57.1	53.6	52.9	- 55.0
Ni	229	115	- 339	243	210	- 279
La	74	55	- 88	38	36	- 39
Ce	142	111	- 162	73	70	- 78
Sm	10.5	9.0	- 11.5	7.1	6.9	- 7.4
Eu	3.1	2.8	- 3.3	2.2	2.1	- 2.3
Tb	1.2	1.0	- 1.5	1.1	0.8	- 1.3
Yb	2.3	2.2	- 2.5	1.6	1.6	- 1.7
Lu	0.32	0.30	- 0.36	0.26	0.24	- 0.28
Hf	7	6	- 8	5	4	- 5

dike	<N>			<O>		
rock type/ number of samples	BA/11			PT/5		

major and minor components, mass fraction in % (wt%)						
	$\bar{x}$	x_{min}	- x_{max}	$\bar{x}$	x_{min}	- x_{max}
SiO_2	40.9	39.4	- 42.4	50.5	43.4	- 55.2
TiO_2	3.7	3.4	- 3.8	2.0	1.2	- 3.0
Al_2O_3	13.2	12.5	- 13.9	17.1	15.5	- 18.5
Fe_2O_3	6.0	4.9	- 7.2	4.2	1.3	- 7.6
FeO	6.0	5.0	- 7.1	3.4	2.1	- 5.5
MnO	0.18	0.13	- 0.21	0.19	0.11	- 0.25
MgO	9.4	7.4	- 11.2	3.3	2.3	- 4.3
CaO	10.4	8.8	- 11.7	5.5	2.0	- 9.1
Na_2O	2.2	1.6	- 3.5	4.4	3.3	- 6.1
K_2O	2.7	1.6	- 3.6	3.1	0.8	- 5.0
ΣH_2O	3.1	2.3	- 4.5	2.8	1.9	- 4.3
P_2O_5	0.67	0.53	- 0.83	0.67	0.36	- 1.1
ΣC as CO_2	0.58	0.10	- 2.4	2.0	0.08	- 4.2
S [$\mu g/g$]	407	270	- 22140	1000	130	- 2860

trace components, μg element/g rock (ppm)						
Sc	30.8	24.1	- 36.6	8.6	3.0	- 14.8
Cr	150	100	- 210	13	n.d.	- 40
Co	49.5	45.8	- 51.9	14.5	1.8	- 31.5
Ni	79	68	- 90	28	n.d.	- 94
La	54	42	- 63	94	31	- 128
Ce	103	81	- 128	172	57	- 243
Sm	10.3	9.0	- 11.6	16.2	6.6	- 22.4
Eu	3.0	2.7	- 3.3	5.0	2.6	- 6.5
Tb	1.2	1.0	- 1.7	1.7	0.8	- 2.8
Yb	2.0	1.5	- 2.3	3.4	1.8	- 4.2
Lu	0.28	0.24	- 0.35	0.40	0.22	- 0.49
Hf	7	4	- 9	11	3	- 18

Table 5.4 CIPW norm for the basalts exposed at the surface.

BA basanite; *ON* olivine nephelinite; *PT* phonolitic tephrite; *nc* not classified basaltic rock

dike	<D>	<H>	<L>		<N>	<O>		<P>
rock type	BA	BA	ON		BA	PT		nc
sample	45/94	43/92	41/90	42/91	44/93	46/87	40/21	46/86
q	-	-	-	-	-	9.7	-	-
or	8.3	14.1	8.3	7.1	9.5	5.0	13.6	8.3
ab	17.8	17.9	1.8	1.9	9.2	34.7	24.4	23.7
an	10.7	13.5	12.2	9.2	16.7	23.3	16.6	18.0
lc	-	-	-	-	-	-	-	-
ne	5.0	6.4	15.6	15.5	11.1	-	8.4	-
di	21.5	19.5	32.0	34.6	28.1	5.3	15.5	23.3
hy	-	-	-	-	-	3.5	-	7.9
ol	13.3	8.0	13.4	16.2	6.7	-	3.0	0.6
he	-	-	-	-	-	6.8	-	-
mt	7.3	5.9	6.4	5.9	7.3	1.2	7.1	8.1
il	5.3	5.9	5.5	4.9	6.8	4.8	5.7	4.8
ap	1.0	1.3	2.6	2.3	1.9	1.1	2.6	1.3
cc	7.5	4.8	0.5	0.3	0.2	0.2	0.7	0.1

Table 5.5 CIPW norm for basalts from the Northern Hessian Depression (WEDEPOHL 1983; calculated from mean values)

	alkali-olivine basalts	basanitic alkali-olivine basalts	nepheline basanites, limburgites	olivine nephelinites
q	-	-	-	-
or	10.6	11.2	10.6	0.9
ab	26.9	12.8	3.6	-
an	16.2	15.5	11.6	11.3
lc	-	-	-	7.2
ne	0.6	6.4	13.6	15.1
di	18.7	24.3	30.3	34.9
hy	-	-	-	-
ol	14.4	16.9	13.9	13.5
he	-	-	-	-
mt	4.5	4.4	5.8	5.9
il	4.2	4.2	5.1	5.1
ap	1.3	1.7	2.0	2.6
cc	-	-	-	-

nents of the basalts studied for this work are depicted graphically in the so-called R_1-R_2 diagram after DE LA ROCHE et al. (1980). The basis for this type of representation is the so-called equivalent numbers (amounts of cations) of the major and minor chemical components (elements) given as oxides.

For the calculation the weight percents of the oxides of the monovalent and trivalent elements are each divided by half of the molar mass. The weight percents of the other components shown were divided by the full molar mass. Following multiplication by 1 000 we have the parameters for Si, Ti, Fe, Na, K, Al, Mg, and Ca, which are, in turn, recalculated to R_1 and R_2 (for detailed explanation see DE LA ROCHE et al 1980). The following example illustrates the calculation using sample 8/17 from the hornblende-bearing basanite of dike <O>:

$M(X)$ = molar mass
w_i = wt% (mass fraction in %, table 5.2)

	$M(X)$	$\dfrac{M(X)}{2}$	w_i	equivalent number
SiO_2	60.1	-	41.6	Si $= \dfrac{41.6}{60.1} \cdot 1\,000 = 692$
TiO_2	79.9	-	3.6	Ti $= \dfrac{3.6}{79.9} \cdot 1\,000 = 45$
Σ Fe as Fe_2O_3	159.7	79.9	13.0	Fe $= \dfrac{13.0}{79.9} \cdot 1\,000 = 163$
Na_2O	62.0	31.0	1.7	Na $= \dfrac{1.7}{31.0} \cdot 1\,000 = 55$
K_2O	94.2	47.1	3.6	K $= \dfrac{3.6}{47.1} \cdot 1\,000 = 76$
Al_2O_3	102.0	51.0	12.9	Al $= \dfrac{12.9}{51.0} \cdot 1\,000 = 253$
MgO	40.3	-	9.0	Mg $= \dfrac{9.0}{40.3} \cdot 1\,000 = 223$
CaO	56.1	-	10.5	Ca $= \dfrac{10.5}{56.1} \cdot 1\,000 = 187$

$$R_1 = 4\,Si - 2\,(Ti + Fe) - 11\,(Na + K) = \mathbf{911}$$
$$R_2 = Al + 2\,Mg + 6\,Ca = \mathbf{1\,821}$$

DE LA ROCHE et al (1980) calculated these parameters for frequent magmatites using more than 25 000 chemical analyses of petrographically classified rocks and compared them with one another. Using frequency distributions they calculated fields for the individual rock types. Figures 5.3 - 5.5 show the results of the calculations for the samples in this study. Since the fields partially overlap due to the smooth transition from one rock type to another, dashed lines were used for indicating the boundaries between them (cf. WIMMENAUER 1985).

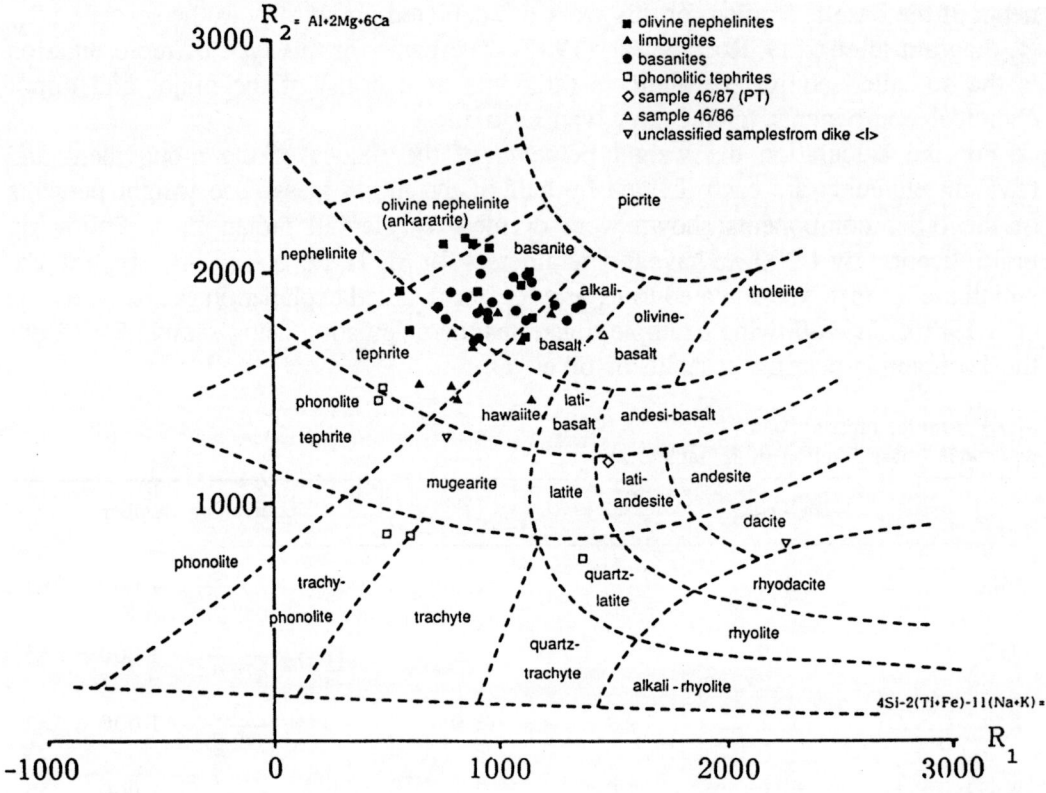

Fig. 5.3 Analytical results of the basalts from the Werra mining district plotted in the R_1-R_2 diagram after DE LA ROCHE et al (1980).

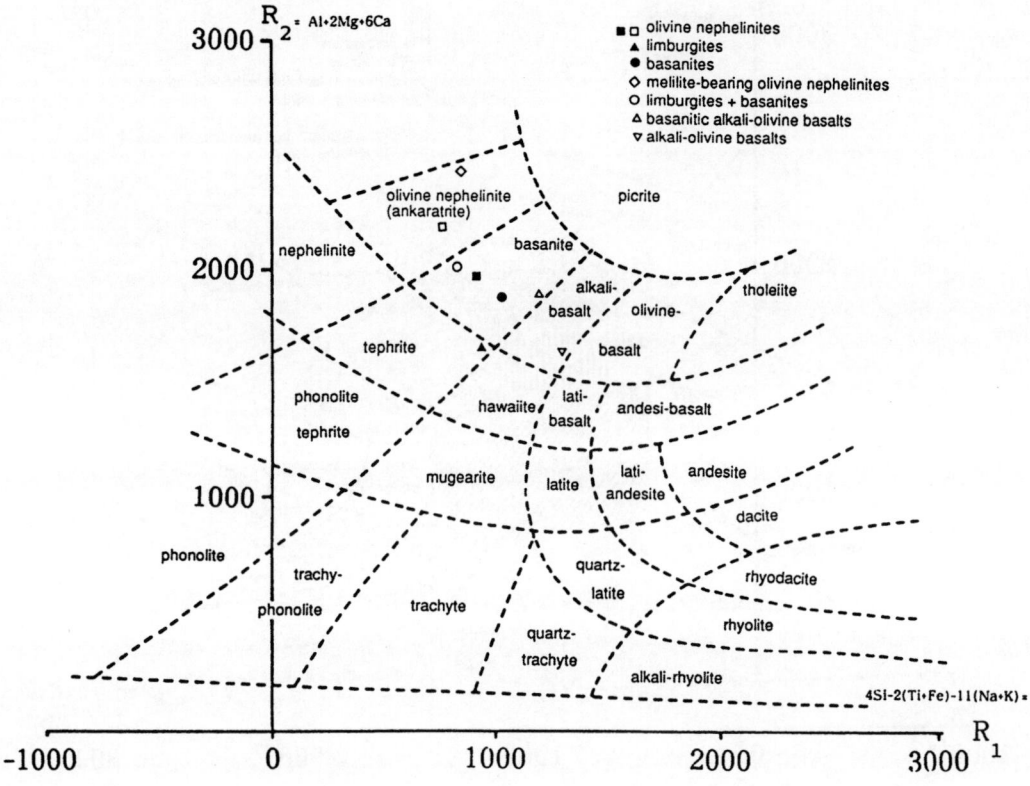

Fig. 5.4 Mean values from the analysis of basalts from the Werra mining district (filled symbols) in the R_1-R_2 diagram (after DE LA ROCHE et al 1980). The R_1-R_2 values calculated from the values of WEDEPOHL (1983) for the mean composition of various basalt types from the Northern Hessian Depression are given for comparision (open symbols). The mean values for the phonolitic tephrite were not calculated due to the great range of variation in R_1-R_2 values (fig. 5.5).

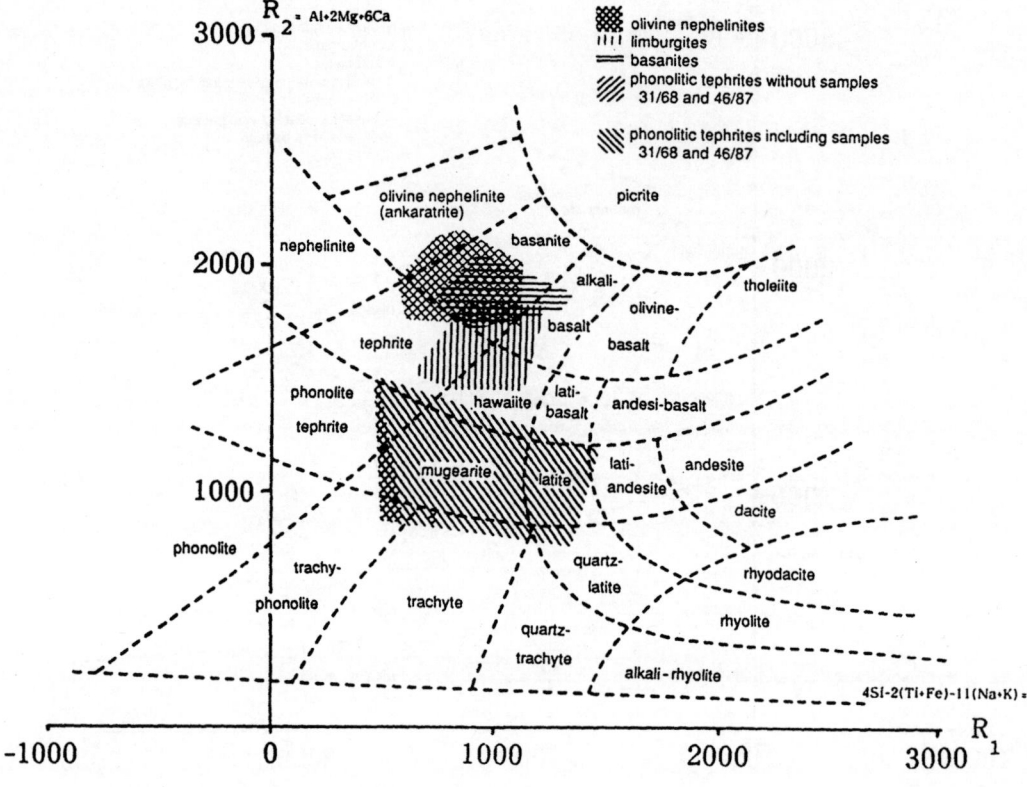

Fig. 5.5 The most extreme points of those shown in fig. 5.3 for one rock type were connected in order to clarify the range of variation in the chemical composition of the different basalts from the Werra mining district (further explanation in text).

The results of the microscopy are largely confirmed when the results of the chemical analyses are depicted in this way. The following statements can be made based on this evaluation:

1. Variations in the chemical composition were already suggested by the differing volume percents of certain minerals observed within the individual basalt dikes (e.g., titanaugite, plagioclase).
2. Similarities in the conditions of formation are underscored by the overlapping areas in the R_1-R_2 diagram covered by the studied olivine nephelinites, limburgites, and basanites (fig. 5.5).
3. Limburgites and basanites are observed to tend toward (basanitic) alkali-olivine basalt. In the case of basanite this is also seen in the mineralogical composition.
4. The limburgites mostly overlap the alkali-olivine basalt, hawaiite (rock similar to tephrite having plagioclase, augite, and olivine as its main components), and tephrite fields. This is an indication that the melts possibly differentiated (fractionation of olivine and clinopyroxenes lowers the R_2 value).
5. The samples microscopically characterized as phonolitic tephrite show the greatest range of variation in chemical composition within one dike. As in the basanites-limburgites, the greatest deviations are found in samples with considerable amounts of microlitic glass here as well (samples 46/87 and 31/68).
6. The not classified sample 46/86 from Dreienberg could be an alkali-olivine basalt based on its chemical composition.

In fig. 5.3 it seems that the autometamorphically altered, microscopically not identifiabled samples can also not be classified using their chemical composition, as expected. The method proposed by DE LA ROCHE et al (1980) also fails when the mineralogical and chemical composition have been extensively altered. It does, however, have the following advantages over the standard CIPW norm for less altered rocks:

- rocks with higher color indices can be described more exactly;
- the comparison of different rock types is simple;
- differences between classifying rocks by R_1-R_2 diagrams and characterizing rocks using microscopy can be evidences of fractionation processes.

Lanthanides

The lanthanide distribution related to chondrites (recommended values for normal chondrites by BOYNTON 1984) in the studied samples shows an enrichment in the lighter elements relative to the heavy ones, as expected (figs. 5.6 - 5.8).

The relationship between the olivine nephelinites, limburgites, and basanites became evident in the absolute amounts of lanthanides as well. Especially the olivine nephelinites and limburgites are very similar. In contrast to the other basalts from the Hattorf mine, the phonolitic tephrite is marked by greater absolute amounts of all lanthanide elements (fig. 5.6).

In fig. 5.7 it is clear that the phonolitic tephrite (dike <O>) contains distinctly greater amounts of lanthanides than sample 46/87, which was assigned to dike <O> due to its mineral constituents. On the other hand, sample 46/87 is very similar to

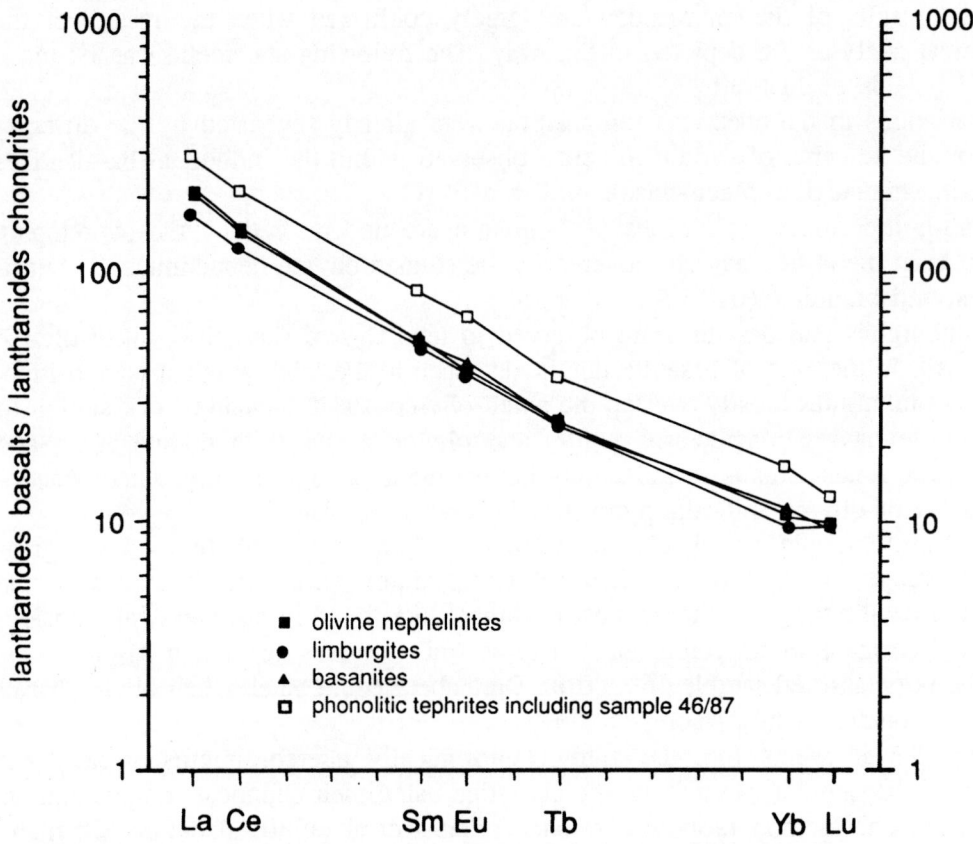

Fig. 5.6 Chondrite-related lanthanide distribution of the basalts from the Hattorf potash mine.

sample 46/86 (not classified) from the same outcrop. These two samples from Dreienberg (outcrop 46-sf) contain smaller amounts of lanthanides than the other basalts (ON, LI, BA).

The distribution patterns of samples 2D/24 and 19/41 from dike <I> lie nearly parallel to one another (fig. 5.8). Due to extreme autometamorphic decomposition of the samples the lanthanide distributions are not comparable to any other basalts in the working area and hence cannot contribute to the classification of rock types.

5.3 Discussion

Although most of the surface and subsurface samples appear macroscopically to be very fresh, their mineralogical and chemical composition indicate that part of these silicate rocks are highly weathered. The composition and genesis of comparable surface specimens from the Northern Hessian Depression west of the Werra mining

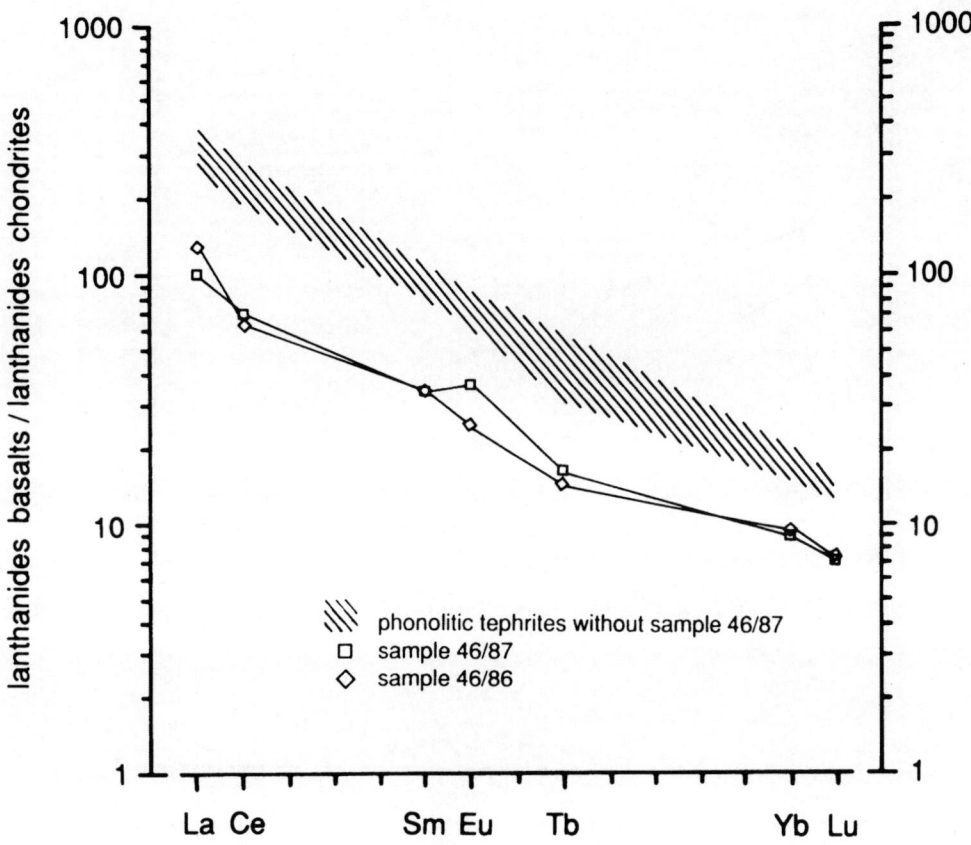

Fig. 5.7 Chondrite-related lanthanide distribution of the phonolitic tephrite (range of variation) and the samples from Dreienberg (outcrop 46-sf).

district are discussed in detail by WEDEPOHL (e.g., 1983, 1985). These rocks will be used for comparison in the following, whereby there are specific differences and distinctive features, particularly in the case of the subsurface samples.

Olivine nephelinite, limburgite, basanite
In tables 5.6 and 5.7 the basanites of this study are listed separately from the limburgites to emphasize the differences between them. The mean values for the two rock types enable comparisons with the data of WEDEPOHL (1985).

SiO_2, Al_2O_3, MnO, CaO. The basaltic rocks only differ gradually in these components and show trends similar to those from the Northern Hessian Depression (table 5.6). Variations are produced by the partially great and variable amounts of water, carbonate, and sulfur in the subsurface basalts and are relativized by standard deviation (used here as a measure of the variability in composition).

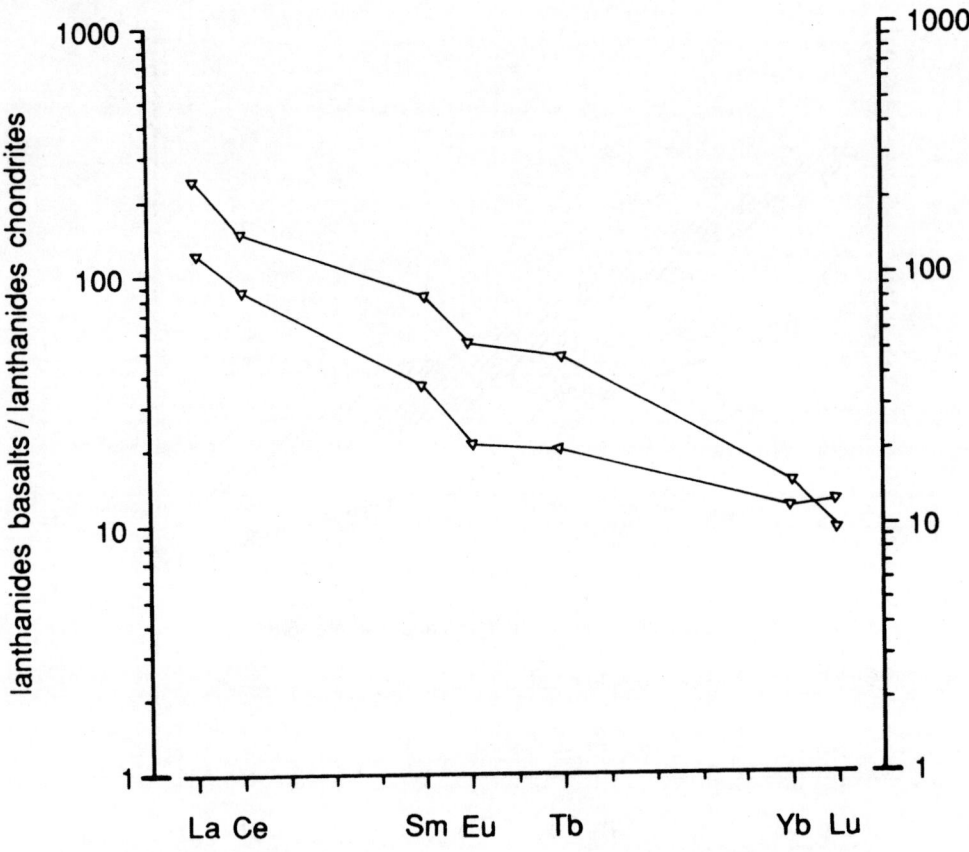

Fig. 5.8 Chondrite-related lanthanide distribution of the unclassified, autometamorphically decomposed samples from dike <I>. The sample with the greatest amount of lanthanides (upper line) is sample 2D/24. The lower line represents sample 19/41.

TiO$_2$, hornblende. In contrast to the rocks from the Northern Hessian Depression, increasingly greater amounts of TiO$_2$ are observed in the olivine nephelinites, limburgites, and basanites. This trend is caused by the great amounts of mafic minerals such as magnetite, augite, and hornblende.

Titanium belongs to the incompatible elements. Due to their physical and chemical properties such elements fit poorly into the structure of minerals occurring in the earth's mantle and consequently are enriched in the magma through partial melting. During crystallization titanium is incorporated preferably into magnetite, clinopyroxenes (augite), and amphiboles.

The formation of amphiboles in magmatic rocks has been investigated by various authors (e.g., AOLI 1963; CAWTHORN 1976; BEST 1974). They report a relationship between pressure and temperature during crystallization of 'basaltic' hornblendes

Table 5.6 Major and minor components in the basalts of the Werra mining district ([1]this study) compared with surface basalts from the Northern Hessian Depression ([2]WEDEPOHL 1985).

$\bar{x}$ mass fraction in % (wt%)

s approximated value of the standard deviation σ (absolute) as a measure of the variation in composition

n.d. no data

rock type	olivine nephelinites[1]			olivine nephelinites[2]			limburgites[1]			basanites[2]		
number of samples	11			9			10			23		
	$\bar{x}$	$\pm$	s	$\bar{x}$	$\pm$	s	$\bar{x}$	$\pm$	s	$\bar{x}$	$\pm$	s
SiO_2	40.1	$\pm$	1.74	0.26	$\pm$	0.95	41.9	$\pm$	1.4	41.3	$\pm$	1.1
TiO_2	2.9	$\pm$	0.32	2.73	$\pm$	0.25	3.5	$\pm$	0.1	3.3	$\pm$	0.4
Al_2O_3	11.7	$\pm$	0.8	11.41	$\pm$	0.78	13.5	$\pm$	1.1	12.7	$\pm$	0.7
Fe_2O_3	4.8	$\pm$	0.6	4.06	$\pm$	0.59	5.8	$\pm$	0.6	5.3	$\pm$	1.0
FeO	6.6	$\pm$	0.4	6.78	$\pm$	0.46	6.0	$\pm$	0.6	6.1	$\pm$	0.6
MnO	0.19	$\pm$	0.01	0.19	$\pm$	0.01	0.19	$\pm$	0.02	0.17	$\pm$	0.02
MgO	11.1	$\pm$	1.7	12.22	$\pm$	2.06	7.5	$\pm$	1.6	9.9	$\pm$	1.4
CaO	11.0	$\pm$	1.0	12.55	$\pm$	0.77	9.4	$\pm$	1.2	10.5	$\pm$	0.9
Na_2O	2.5	$\pm$	0.8	3.34	$\pm$	0.65	2.1	$\pm$	0.8	2.2	$\pm$	0.6
K_2O	2.0	$\pm$	0.5	1.70	$\pm$	0.49	3.1	$\pm$	0.4	2.4	$\pm$	0.6
ΣH_2O	3.7	$\pm$	1.1	2.7		-	4.5	$\pm$	1.0	3.4	$\pm$	0.8
P_2O_5	0.8	$\pm$	0.2	0.88	$\pm$	0.18	0.78	$\pm$	0.07	0.6	$\pm$	0.1
ΣC as CO_2	1.7	$\pm$	2.0	n.d.		-	0.9	$\pm$	0.4	1.3	$\pm$	1.2

rock type	basanites + limburgites[1]			nepheline basanites + limburgites[2]			basanitic alkali-olivine basalts[2]			phonolitic tephrites[1]		
number of samples	33			11			13			6		
	$\bar{x}$	$\pm$	s	$\bar{x}$	$\pm$	s	$\bar{x}$	$\pm$	s	$\bar{x}$	$\pm$	s
SiO_2	41.5	$\pm$	1.2	42.27	$\pm$	1.17	45.18	$\pm$	1.13	50.5	$\pm$	4.3
TiO_2	3.4	$\pm$	0.3	2.65	$\pm$	0.33	2.23	$\pm$	0.12	2.0	$\pm$	0.8
Al_2O_3	13.0	$\pm$	0.9	11.77	$\pm$	0.56	12.47	$\pm$	0.95	17.1	$\pm$	1.11
Fe_2O_3	5.4	$\pm$	0.9	4.02	$\pm$	0.94	3.00	$\pm$	0.54	4.2	$\pm$	2.1
FeO	6.0	$\pm$	0.6	6.85	$\pm$	0.73	7.49	$\pm$	0.59	3.4	$\pm$	1.2
MnO	0.18	$\pm$	0.02	0.17	$\pm$	0.01	0.18	$\pm$	0.02	0.19	$\pm$	0.05
MgO	9.2	$\pm$	1.8	11.46	$\pm$	1.04	10.95	$\pm$	2.45	3.3	$\pm$	0.8
CaO	10.2	$\pm$	1.1	11.22	$\pm$	1.03	10.18	$\pm$	0.59	5.5	$\pm$	2.8
Na_2O	2.1	$\pm$	0.7	3.38	$\pm$	0.83	2.88	$\pm$	0.57	4.4	$\pm$	0.9
K_2O	2.6	$\pm$	0.6	1.81	$\pm$	1.12	1.89	$\pm$	0.15	3.1	$\pm$	1.3
ΣH_2O	3.7	$\pm$	1.0	2.3		-	2.1		-	2.8	$\pm$	1.0
P_2O_5	0.7	$\pm$	0.1	0.88	$\pm$	0.18	0.73	$\pm$	0.16	0.7	$\pm$	0.3
ΣC as CO_2	1.2	$\pm$	1.0	n.d.		-	n.d.		-	2.0	$\pm$	1.8

Table 5.7 Trace components in the basalts of the Werra mining district ([1]this study) compared with surface basalts from the Northern Hessian Depression ([2]WEDEPOHL 1985).

$\overline{x}$ mass fraction in µg element/g rock (ppm)

s approximated value of the standard deviation σ (absolute) as a measure of the variation in composition

rock type	olivine nephelinites[1]		olivine nephelinites[2]		limburgites[1]		basanites[2]	
number of samples	11		5 to 8		10		23	
	$\overline{x}$ ± s		$\overline{x}$ ± s		$\overline{x}$ ± s		$\overline{x}$ ± s	
S	1930	±2550	247	± 54	1740	± 670	2800	± 4460
Sc	27.4 ±	2.1	21 ±	1.7	22.4 ±	3.4	29.9 ±	3.2
Cr	360 ±	120	371	± 100	70 ±	80	240 ±	110
Co	52.0 ±	4.1	49 ±	5	40.8 ±	6.6	50.2 ±	4.1
Ni	205 ±	78	329	± 220	55 ±	32	128 ±	64
La	65 ±	13	112 ±	11	65 ±	3	51 ±	9
Ce	121 ±	22	217 ±	21	121 ±	8	98 ±	18
Sm	10.0 ±	1.1	13.7 ±	1.3	10.4 ±	0.9	9.3 ±	1.4
Eu	0.9 ±	0.2	3.9 ±	0.6	3.2 ±	0.1	2.8 ±	0.4
Tb	1.2 ±	0.2	1.6 ±	0.3	1.1 ±	0.3	1.1 ±	0.3
Yb	2.1 ±	0.2	2.0 ±	0.65	2.2 ±	0.1	1.9 ±	0.3
Lu	0.30±	0.03	0.28 ±	0.04	0.30±	0.03	0.28±	0.03
Hf	6 ±	1	7.0 ±	0.7	7 ±	2	6 ±	1

rock type	basanites + limburgites[1]		nepheline basanites + limburgites[2]		basanitic alkali-olivine basalts[2]		phonolitic tephrites[1]	
number of samples	33		2 to 12		11 to 14		6	
	$\overline{x}$ ± s		$\overline{x}$ ± s		$\overline{x}$ ± s		$\overline{x}$ ± s	
S	2480	±3770	363	± 452	103	± 40	1000	± 950
Sc	27.7 ±	4.7	20 ±	2.8	19.9 ±	0.6	8.6 ±	5.5
Cr	190 ±	130	397 ±	23	395	±206	10 ±	10
Co	47.3 ±	6.6	52 ±	3.9	49 ±	7.3	14.5 ±	12.4
Ni	106 ±	65	334	±126	301	±242	28 ±	31
La	55 ±	10	82 ±	4	63 ±	13	94 ±	32
Ce	105 ±	19	151 ±	14	135 ±	24	172 ±	58
Sm	9.6 ±	1.3	12 ±	0.9	9.2 ±	1.1	16.2 ±	4.9
Eu	2.9 ±	0.4	3.4 ±	0.3	2.8 ±	0.3	5.0 ±	1.2
Tb	1.1 ±	0.3	1.2 ±	0.3	0.98 ±	0.14	1.7 ±	0.6
Yb	2.0 ±	0.3	1.9 ±	0.3	1.9 ±	0.35	3.4 ±	0.8
Lu	0.28±	0.03	0.29 ±	0.02	0.27 ±	0.05	0.40±	0.08
Hf	6 ±	2	6,1 ±	2	4.8 ±	0.8	11 ±	5

and the amount of TiO_2 and Fe/Mg ratio of the crystals. For example, titanium is preferably incorporated into the amphibole structure with decreasing pressure. An increase in the Fe/Mg ratio in crystallizing hornblendes is observed with the cooling of the silicate melt. Hornblendes from the mantle also contain greater amounts of Cr than amphiboles which crystallize out of basaltic magma.

Of the rocks used for comparison only one outcrop was composed of hornblende-bearing basalt, i.e. the basanite of Rosenberg northeast of Kassel. Tuffs are common to this occurrence and surface outcrop 44-of of dike <N>. An explosive eruption of these magmas can be inferred from this. Unlike the Rosenberg basanite with about 0.8 vol% pargasitic-kaersutitic hornblendes (VINX & JUNG 1977), the basanite of dike <N> contains up to 25 vol% hornblende. The Rosenberg basanite contains amphibole-bearing peridotite xenoliths and aggregates of hornblende crystals. The basanite of dike <N>, however, contains exclusively idiomorphic hornblende crystals (pargasite-hastingsite). The hornblendes from five outcrops of dike <N> were analyzed with the microprobe for an initial genetic interpretation. The results are compiled in table 5.8 together with analyses of the Rosenberg basanite and of peridotite inclusions from the same locality (VINX & JUNG 1977). The mean values of the hornblendes in peridotite xenoliths from the melilite-bearing olivine nephelinite of Westberg near Hofgeismar (also Northern Hessian Depression) analyzed by OEHM (1980) are also given. Finally, table 5.8 lists the chemical composition of hornblendes from peridotites based on a compilation of data from WEDEPOHL (1975).

As expected, the analyzed amphiboles differ from each other primarily in their TiO_2 and Cr_2O_3 contents as well as their Fe/Mg ratio. The hornblende from the basanite in the Werra mining district obviously crystallized out of the silicate melt at lower temperatures and pressures. The amphiboles from the mantle and the hornblendes which crystallized in subcrustal levels contain distinctly lower amounts of TiO_2 and relatively high amounts of Cr_2O_3. In addition, the Fe/Mg ratio is characteristically lower than that of the amphiboles from dike <N> (table 5.8). Since the basanite of the Hattorf mine is hosted by evaporites, its Na_2O and K_2O contents were not interpreted.

The chemical composition of the hornblendes frequently varies depending upon the pressure and temperature during crystallization, also observed by VINX & JUNG (1977). The hornblendes from dike <N>, on the other hand, have a very uniform composition and are not zoned at all. There were also no determinable significant differences between crystals of different size.

Since amphiboles are no longer in equilibrium with the silicate melt during a decrease in pressure, they easily decompose to opacite, which here consists of a dense mass of augite, magnetite, rhoenite, olivine, plagioclase, and brown glass (TRÖGER 1967). The crystals from the basanite of dike <N>, however, are always very well preserved and only have a very fine rim of opacite. This is a result of the sudden drop in temperature chilling the magma during intrusion. The great amount of glass in the groundmass can also be explained in this way.

Table 5.8   Microprobe analyses of hornblendes of dike <N> compared with hornblendes of basanite and peridotite inclusions from Rosenberg (VINX & JUNG 1977) and with hornblendes of peridotite xenoliths from Westberg (OEHM 1980) and a summary of WEDEPOHL (1975).

n   number of analyzed mineral grains; the standard deviation given here is a measure of the variability in the chemical composition of the hornblendes of dike <N> (minimum of three measurements per grain).

n.d. no data
ud. undetectable

	dike <N> basanite n = 23	Rosenberg basanite n = 2	Rosenberg peridotite n = 2	Westberg peridotite n = 3	- peridotite n.d.
SiO_2	40.6 ± 0.6	40.7	43.0	42.7	43.6
TiO_2	4.4 ± 0.3	4.7	3.6	1.6	0.7
Al_2O_3	14.3 ± 0.4	14.1	13.1	15.0	13.2
FeO	10.7 ± 0.9	9.5	5.5	4.5	4.0
MgO	13.6 ± 0.6	13.9	17.7	17.9	19.2
CaO	11.2 ± 0.8	11.2	11.0	10.9	11.2
Na_2O	2.2 ± 0.1	2.6	2.9	3.3	3.4
K_2O	1.4 ± 0.3	2.1	1.5	0.8	0.7
Cr_2O_3	ud.	ud.	0.9	0.7	1.9
$\dfrac{Fe}{Mg}$	1.01	0.88	0.40	0.32	0.27

A greater amount of volatile components (H_2O, CO_2) is generally necessary in mantle rocks for the formation of SiO_2-undersaturated melts (e.g., BREY & GREEN 1975; EGGLER 1974). The crystallization of hornblendes (and biotite) out of basaltic melts, particularly in the observed 'pegmatitic' form, is evidence of high hydrostatic pressure. The extreme amounts of fluid phases are also indicated by the occurrence of native sulfur at the contact between the basanite and the salt rock (chapter 6). In this context it is noteworthy that the hornblende-bearing basanite from Rosenberg with 884 μg S/g rock (ppm) has the highest sulfur content of the surface basalts of the Northern Hessian Depression discussed by WEDEPOHL (1983).

As the magma ascended from the mantle, it appears to have been enriched in volatile components (magmatic differentiation). This does not conflict with the rapid production of melts discussed in chapter 9. The distinct effects of decomposition of the early crystallized pyroxene phenocrysts are an indication of magmatic differentiation. The aforementioned uniform composition of the amphiboles from the Hattorf mine was possibly caused by a more or less simultaneous crystallization at a relatively shallow depth. The genesis of the amphiboles will be elucidated in a forthcoming paper (FREERK 1990, in preparation).

In comparison to dike <N> (basanite), hornblendes in the rocks from dike <I> (not classified) and <O> (phonolitic tephrite) occur more seldom and are not as well developed. The pyroxenes in dike <I> were not converted into amphiboles as in the phonolitic tephrite of dike <O>.

Fe$_2$O$_3$, FeO. In contrast to the rocks used for comparison, the Fe(III)/Fe(II) ratio in the studied basalts clearly increases from the olivine nephelinites to the limburgites and basanites. In spite of the rapid cooling of the limburgites (large amounts of glass), the original Fe(II) was obviously not fixed and did oxidize into Fe(III) through interaction with fluid phases.

The surface samples of the various basalt types have a very variable Fe(III)/Fe(II) ratio. Here, the original ratio has possibly changed due to weathering, i.e. oxidation of the Fe(II).

MgO, olivine. Smaller fragments of peridotite were only observed in the olivine nephelinites. The amounts of Mg which were significantly greater in the olivine nephelinites (especially in surface samples) compared with the basanites and limburgites can be explained by the assimilation of such peridotite xenoliths during ascent of the magma. In addition, the basanitic melts were slightly fractionated through the separation of olivine (detailed discussion below).

In all investigated dikes, olivine has been altered to a partially yellow, fibrous substance along cracks and breaks in the crystals. This mineral is chrysotile and forms by absorption of H$_2$O and depletion of magnesium. The magnesium is fixed in neogenic carbonate minerals or Mg-phyllosilicates (KNIPPING 1984) .

Na$_2$O, K$_2$O. The weight percents of these components in the subsurface samples vary greatly within the individual dikes as well. The data in table 5.6 show that the mean values for Na$_2$O trend from higher concentrations in the olivine nephelinites toward lower ones in the limburgites and basanites. The opposite is the case for the K$_2$O concentrations. It must be pointed out here that the surface samples with 'normal' amounts of alkali elements were also included in the calculation of the mean value for the olivine nephelinites and basanites.

As can be seen in the data of the rocks used for comparison (table 5.6), no sodium was absorbed by the basaltic melts in the Werra region during intrusion into the Zechstein evaporites. There is, however, an increase in the amounts of potassium. This potassium is fixed in neogenic phyllosilicates. This has already been described several times for basaltic rocks in evaporites (e.g. WIMMENAUER 1952, KOCH & VOGEL 1980, KNIPPING 1984).

The depletion of sodium and enrichment of potassium are greatest in the limburgites, compared with the other basalts from the Hattorf mine (table 5.6). The sodium in the nepheline of the glassy groundmass was apparently lost. The formation of microlites (devitrification) led to the fixing of potassium. Due to these great changes in the initial amounts of Na and K these elements cannot be used in a genetic interpretation.

Σ H$_2$O, Σ C as CO$_2$. The amounts of H$_2$O in the subsurface sample were higher than in the samples used for comparison, as expected. This H$_2$O is fixed in biotite, neogenic phyllosilicates, analcime, amphiboles, and particularly in glass. In contrast, the basalts exposed at the surface in the area of the Hattorf mine contain amounts of H$_2$O that are similar to the basalts west of the Werra region.

Elevated amounts of carbonate in surface basanites 43-sf and 45-sf are attributed to weathering (table 5.2). The portions of carbonate in most subsurface samples were unusually high as well. The carbonates formed by decomposition of silicates in a CO$_2$-bearing phase and are generally indicative of the intensity of the alteration process. For example, the autometamorphically decomposed, unclassified samples from dike <I> contain 9.5 wt% C (sample 2D/24) and 13.6 wt% C (sample 19/41) as CO$_2$. According to KNIPPING (1984) the microcrystalline carbonates in outcrop 2D/24 contain an average of 10 wt% FeCO$_3$, 50 wt% CaCO$_3$, and 40 wt% MgCO$_3$, in addition to traces of MnCO$_3$. The ratio between the various cations can vary greatly in the carbonate compounds of one sample.

The amount of carbonate in the olivine nephelinites, which is larger on the average than in the basanites and limburgites, could be attributed to larger amounts of olivine in the former and thus to a greater availability of magnesium.

Sulfur. A detailed discussion on sulfur follows in chapter 6.

Sc, Hf. With a distribution coefficient near 1 scandium has hardly been fractionated during partial melting. Compared with a 'depleted' mantle rock (WEDEPOHL 1981) the enrichment factor in the studied rocks accordingly amounts to only about 3.5. The incompatible element hafnium, contrarily, was enriched by a factor of 20.

Co, Ni, Cr. These so-called residual elements fit well into the crystal structures of mantle minerals due to their crystal-chemical properties. The amounts of these elements vary greatly in the basalt samples (tables 5.2, 5.3). The amounts of cobalt, nickel, and chrome in the studied olivine nephelinites are comparable with the corresponding basalts west of the Werra region (table 5.7). As in the case of magnesium, the high contents in sample 42/91 from Soisberg (ON), for example, could be attributed to assimilated peridotite xenoliths.

Olivine and augite have the highest distribution coefficients for nickel and chrome of all the possible basalt minerals. As a basaltic melt crystallizes the magnesium in olivine is partially replaced by nickel. The Co/Mg ratio in magmatic rocks is relatively constant due to the incorporation of cobalt in the early precipitated Mg minerals (olivine; e.g., NOCKOLDS & ALLEN 1956). Yet, in contrast to nickel, the selective enrichment of cobalt in olivine is less distinct.

Chromium is preferably incorporated by the basaltic clinopyroxenes of a melt. The residual trace elements are relatively resistent to alteration, e.g., the lanthanides (e.g., MOTTL & HOLLAND 1978; MOTTL et al 1979; SEYFRIED & MOTTL 1982; MENGEL et al 1987). Hence, the low Ni and Cr contents in the basanites and limburgites indicate the fractionation of olivine and pyroxene out of a melt.

Lanthanides. The affinity between the analyzed rocks becomes evident in a comparison of the chondrite-normalized values for the lanthanides (chapter 5.2). This is true above all for the olivine nephelinites and limburgites. It could indicate that the two rocks can be classified together as the same basalt type. Lanthanides are, of course, enriched in the melt during magmatic differentiation (e.g., HASKINS 1984), but they do not show any or just a slight tendency to fractionate during alteration processes (e.g., HAJASH 1984). It can therefore be concluded that the differentiated limburgites (see below) have been enriched in lanthanides, and that their present distribution does not correspond to that in the initial melt.

The absolute amounts primarily of the light lanthanides in the basalts from the Hattorf mine are somewhat smaller in relation to the rocks used for comparison (table 5.7). Since the enrichment of incompatible lanthanides in a basaltic magma is dependent upon the degree of dissolution, this finding emphasizes the intermediate character of the investigated basalts.

Phonolitic tehprite
The rock of dike <O> described as phonolitic tephrite based on its mineralogy also differs greatly from the other investigated rocks and rocks used for comparison in terms of their chemical composition.

Variations both in mineralogy and chemical composition are additionally found over the NS extension of the dike (chapters 5.1, 5.2). The differences in the SiO_2 and CaO contents as well as trace element contents are especially evident. The latter is true above all for sample 46/87, the mineralogy of which, however, does not resemble that of sample 31/68.

There are various indications that the phonolitic tephrite is a differentiated rock with distinct fractionation primarily of early crystallized olivines and clinopyroxenes. In comparison to the previously discussed basaltic rocks (ON, LI, BA) the following trends have been observed for the phonolitic tephrites (the constituents in parentheses to a limited extent):
- enrichment in SiO_2, Al_2O_3, Na_2O, (K_2O), (P_2O_5), Hf and the lanthanides,
- depletion in (TiO_2), (Fe_2O_3), FeO, MgO, (CaO), Sc, Cr, Co, and Ni.
Two groups can clearly be distinguished, as already suggested by the petrographic studies. The degree of enrichment or depletion is dependent upon whether hornblende and/or pyroxene occur in the rock and only olivine was fractionated (samples 46/87, 40/21, and 33/723), or whether nearly all mafic minerals are lacking in the rock (samples 50/85, 12/25, and 31/68). The relatively high Na contents also in surface sample 46/87 can be explained by the high volume percents of plagioclase and nepheline. Potassium is also fixed in potassium feldspar, in addition to phyllosilicates.

The phonolitic tephrite from Dreienberg near Friedewald (sample 46/87) is considerably depleted in Mg, in spite of the presence of relictic olivine. The latter, however, has been converted completely into iddingsite (aggreate of mainly goethite with hematite and submicroscopic clay minerals), as seen under the microscope. This conversion by a hydrothermal phase results in a depletion in Mg and oxidation of Fe(II) (the sample has a Fe_2O_3/FeO ratio of 3). In fact, the Ni contents, for example, also indicate substantially less olivine fractionation in this rock as compared with the other samples from dike <O>.

Regarding lanthanides, sample 46/87, which was described as phonolitic tephrite, is very similar to the unclassified rock of sample 46/86 (fig. 5.8). Both samples from surface outcrop 46-sf are substantially less enriched in chondrites than the other phonolitic tephrites. The rock from outcrop 46-sf was described as feldspar basalt by the first workers on sheet Friedewald (no. 5125; v.KOENEN 1888, BÜCKING 1881). This term corresponds to a alkali-olivine basalt in present-day basalt nomenclature. This weathered basalt is in fact comparable with alkali-olivine basalts with respect to the still recognizable mineral constituents and the major and minor elements (fig. 5.3; WEDEPOHL 1983). The phonolitic tephrites of dike <O> are possibly a product of the differentiation of an alkali-olivine-basaltic parent magma, especially since the glass in this basalt type has a phonolitic composition (GRAMSE 1971).

The phonolitic tephrite of dike <O> can be separated into two groups. One group contains hornblendes and pyroxenes, whereas in the other group these minerals are completely absent. Hence, it can be concluded that dike <O> possibly represents two, more or less independent fracture systems or intrusions. Yet, since there is no conclusive evidence for this, both samples were assigned to dike <O>.

No relatively fresh samples of the tephrite-like rock of dike <I> were collected for lack of outcrops. Consequently, this dike was not able to be described in more detail.

Fractionated magmatic differentiation
In recent years, knowledge concerning the mechanism of magma ascent from the mantle has been expanded greatly. In addition, the velocities of the melts up to the surface have been determined experimentally (e.g., CARMICHAEL et al 1977; KUSHIRO et al 1976; SCARFE et al 1980; SPERA 1980). The presence of relatively heavy peridotite xenoliths in the basalts accordingly indicate that the magma ascended from the mantle to the surface within hours or days. This and other evidence (see above) suggest that of the silicate rocks investigated here at least the olivine nephelinites could be products of unfractionated partial melts from the mantle.

No surface samples were able to be used for gathering reliable information on the extent of fractionated magmatic differentiation. The limburgites namely show no sign of magma extrusion at the surface. Moreover, balancing the amounts of fractionation is extremely difficult because the composition of the investigated subsurface samples was influenced by the host rocks (evaporites) and the site of their exposure (nearly 1 000 m below the surface with the overburden hindering de-gassing).

Clear evidence of whether a basalt has crystallized out of an unaltered partial melt from the mantle is, for example, the following (WEDEPOHL 1985; MENGEL et al 1987):
- the Mg/Fe(II) ratio or so-called Mg number = 100 Mg/Mg+Fe(II),
- the Ni/Co ratio in basalt, which is dependent upon olivine fractionation, and
- a Rayleigh fractionation model for closed systems.

The initial weight percent of the Fe(II) in a basaltic melt is included in the Mg number. However, the analytical results obtained in this study show a very strong oxidation of Fe(II) into Fe(III) in several samples. Yet, recalculating the original amount of Fe(II) using the relatively constant Cr/Fe(III) ratio in basaltic melts appears to be too inaccurate due to the highly variable amounts of chrome in the studied samples.

Another possibility for correcting the Fe(II) value is to assume a constant Fe(III)/ Σ Fe ratio (e.g., FREY et al 1978; WEDEPOHL 1983). Greater quantities of fluid phases (H_2O, CO_2) are necessary for generating highly SiO_2 depleted olivine nephelinite melts or hornblende-bearing basanites (e.g., GREEN 1972; BREY and Green 1975). According to WEDEPOHL (1983), however, increased oxygen fugacity caused by this can elevate an otherwise constant Fe(III)/Σ Fe ratio in dependency upon the degree of dissolution of the mantle material. For these reasons, calculating the Mg/Fe(II) ratio does not appear expedient.

Since the element nickel is fixed preferably in early segregated olivine during crystallization of a basaltic melt, the Ni/Co ratio of SiO_2-depleted basalts can be an indication of possible olivine fractionation. Whereas the corresponding rocks used for comparison had Ni/Co ratios of 6.1 - 6.7 (WEDEPOHL 1985, p. 130), the silicate rocks studied here from the Werra mining district had the following Ni/Co values:

olivine nephelinites	3.9
sample 42/91 (ON from Soisberg)	5.9
limburgites	1.3
basanites	2.5
phonolitic tephrite (without sample 46/87)	1.3
sample 46/87 (phonolitic tephrite)	3.0
sample 46/86 (potential alkali-olivine basalt)	3.8

These results indicate an olivine fractionation in all studied rocks, excluding the peridotite-bearing olivine nephelinite from Soisberg (sample 42/91). Fractionation was greatest in the cases of phonolitic tephrites and limburgites.

The following calculation of balance should give a simple impression of the amount of fractionation in the early precipitates of olivine and clinopyroxene. A Rayleigh fractionation model for closed systems will be used for this purpose. The calculations are based on the weight percents of nickel and chrome in the investigated rocks. As mentioned above, olivine preferably incorporates nickel. No inclusions of chrome-bearing minerals (e.g., Cr-spinell; WEDEPOHL 1963) were observed in the olivine crystals. Hence, a depletion of chrome in basalts can be attributed to the fractionation of clinopyroxenes. The following mathematical relationship is valid (e.g., HART & ALLÉGRE 1980):

$$C_1 = C_0 \cdot f^{(D-1)} \tag{5.1}$$

C_1 concentration of the element in the melt (mass fraction in $\mu g/g$ determined in the basalt),

C_0 concentration of the element in the parent magma or in an unfractionated rock of the corresponding magma type (a basanitic parent magma is assumed for the limburgites, and an alkali-olivine basaltic parent magma for the phonolitic tephrite),

f portion of melt to be calculated after fractionation; $f = 1$ is true for an unfractionated melt,

D distribution coefficient of the element between mineral (olivine or pyroxene) and the melt; these values were taken from the data of WEDEPOHL (1985, p. 132).

By transforming equation (5.1) we obtain

$$\frac{\lg \frac{C_1}{C_0}}{D - 1} = \lg f \qquad (5.2)$$

Table 5.7 shows the calculated results for the various basalts and for samples 46/86 (not classified, potential alkali-olivine basalt) and 46/87 (phonolitic tephrite from Dreienberg, outcrop 46-sf). In the case of the phonolitic tephrite about 40 wt% of the parent magma was fractionated as olivine and clinopyroxene.

Limburgites are actually vitrified (glassy) basanites (hyalobasanites) and hence undifferentiated rocks whose primary composition has not been changed. This is evidenced by the peridotite xenoliths frequently observed in limburgites. Regarding their chemistry the rocks termed limburgites studied here can be characterized as transitional forms between basanites, hawaiites, and tephrites. This is also demonstrated by the fact that the limburgites also appear to have lost a major portion of the initial melt. Nevertheless, these rocks contain high amounts of glass. Thus, the petrographical classification of these rocks was greatly hindered and so they have been designated as limburgites (see, e.g., Cox et al 1979). The amounts of titanium in the limburgites, which are higher compared with the basanites, could possibly be explained by an enrichment of this element in a Fe-Ti-oxide phase.

The basanites are not as strongly differentiated. The same is true for sample 46/87, which was described as a phonolitic tephrite due to its mineralogical composition, and above all for the potential alkali-olivine basalt (sample 46/86). With an average of about 5 wt% of the parent magma, which was fractionated as olivine and pyroxene, the olivine nephelinite was the least differentiated rock in the study area. The elevated amounts of Ni and Cr in the olivine nephelinite from Soisberg (sample 42/91) are attributed to assimilated fragments of peridotite.

Regarding the accuracy of the results it should be mentioned that the calculations involved a simplified balancing. The data for C_1 have possibly been falsified by the elevated H_2O, CO_2 and S contents in the subsurface basalts. It must also be presumed that other minerals like magnetite, for example, have been fractionated besides olivine and pyroxene. Since, however, the model calculations confirm the observations from microscopy and the previous interpretation of the composition these results may be realistic in terms of their magnitude.

The depth at which fractionation occurred was not able to be determined in this study. Assuming that the rock of dike <O> also originated from the potentially alkali-olivine basaltic parent magma of sample 46/86, fractionation must have occurred, however, at greater depths because the two rocks have been magnetized in opposite

Table 5.9 Simplified model calculation of the olivine and clinopyroxene fractionation in the basaltic rocks of the Werra region.

The Ni and Cr contents for rocks that crystallized out of parent primary mantle material (C_1 and C_0 in µg/g) were taken from WEDEPOHL (1985, tables 4a and 3). The Cr contents for primary alkali-olivine basalts were calculated from the data of Wedepohl (1983, table 13; 1985, table 4a).

ON olivine nephelinite; *LI* limburgite; *BA* basanite; *PT* phonolitic tephrite without sample 46/87; samples 46/87 and 46/86 from Dreienberg near Friedewald.

PT^1 with relictic pyroxenes and hornblendes (samples 40/21, 33/723)

PT^2 without pyroxenes and hornblendes (samples 50/85, 12/25, 31/68)

$ol \cdot 100$, $cpx \cdot 100$, mass fraction in % (wt%) of the parent magma which was fractionated as olivine and clinopyroxene, respectively

$\Sigma f_l = 100 - (ol \cdot 100 + cpx \cdot 100)$, rounded-off value for the differentiated mass fraction in % of a parent magma, which is the basalt of today

	ON	LI	BA	PT^1	PT^2	46/87	46/86
$C_{1,Ni}$	205	55	128	15	14	94	173
$C_{0,Ni}$	329	329	334	308	308	308	308
$D_{ol,Ni}$	10	10	10	10	10	10	10
f_{ol}	0.949	0.820	0.899	0.715	0.709	0.876	0.938
$ol \cdot 100$	5.1	18	10.1	28.5	29.1	12.4	6.2
$C_{1,Cr}$	360	70	240	15	3	40	210
$C_{0,Cr}$	371	371	397	439	439	439	439
$D_{cpx,Cr}$	34	34	34	34	34	34	34
f_{cpx}	0.999	0.951	0.985	0.903	0.860	0.930	0.978
$cpx \cdot 100$	0.1	4.9	1.5	9.7	14.0	7.0	2.2
Σf_l	95	77	89	62	57	81	92

directions, as mentioned in chapter 5.1 (SIEMENS 1971). There is an average of about 300 000 to 500 000 years between reversals in the earth's magnetic field. Within this time span a primary magma which intruded into shallow depths should have already cooled and solidified long before a subsequent intrusive event.

Amygdules, ocelli

Halite-filled amygdules and aggregates were observed only in the subsurface basalts. None of the groundmass minerals of the basalts have acted as fillings in the amygdules and vesicles. In contrast, the clusters and ocelli are distinguished by a smooth transition between the filling and groundmass of the basalt.

The ocelli- and amygdules-bearing basalts in the potash salt occurrences of Buggingen have been discussed in detail by HURRLE (1976). The formation temperatures of the aggregates were higher than those of the amygdules. In the basalts from the Hattorf mine amygdules occur above all in glass-rich portions of the rock and in the limburgites. They can thus be interpreted as gas vesicles which were filled after solidification of the basaltic melts. At the margins the fillings consist of neogenic carbonates formed through the decomposition of the silicate minerals. Further inward, idiomorphic anhydrite crystallized. The bulk of the filling is halite, which formed last.

According to HURRLE (1976) the frequent tangential orientation of groundmass minerals at the margin of the ocelli shows that they formed before the complete solidification of the silicate melt. Generally, the occurrence of amygdules and ocelli in basalts evidences the presence of larger quantities of volatile phases as the melt cooled. Hence, this is a frequent phenomenon especially in dike rocks. This is illustrated, for example, by the biotite-rich ocelli fillings in the olivine nephelinites from the Hattorf mine. The presence of halite with gas and fluid inclusions in the amygdules and aggregates attests to the fact that the aqueous phases transported in the silicate melts have been obviously capable of absorbing NaCl during ascent through the evaporites.

The formation of ocelli through the incomplete assimilation of rock salt xenoliths is surely not a dominant process. The ocelli are always round. In addition, the idiomorphic anhydrite has grown into the amygdules and ocelli before or at the same time as the halite crystallized.

KOCH (1978) and KOCH & VOGEL (1980) presume that the amygdules and ocelli were produced by the exsolution of rock salt xenoliths dissolved in the magma. The rock salt considered here could be described in a simplified way by the NaCl - $CaSO_4$ system. Like the KCl - $CaSO_4$ system (e.g., HEIDE & BRÜCKNER 1967), the NaCl - $CaSO_4$ system is a eutectic system lacking solid solution (LEVIN et al 1964, cited in RAWSON et al 1967). With up to 35 % $CaSO_4$, halite first crystallizes out of the melted rock salt fragments during cooling. At the eutectic point (725 °C) $CaSO_4$ then crystallizes as well. In contrast, the crystallization sequence is just the opposite in the studied amygdules and ocelli from Hattorf, as is expected for the formation from aqueous solutions. When a melt cools rapidly or chills (e.g., near the contact to the host rock), perthite-like exsolution structures would have to form according RAWSON et al (1967, fig. 6). Such structures were not observed in any of the basalt thin sections studied.

Volatile components

As explained above, there is much evidence, contrary to KOCH & VOGEL (1980), that fluid phases (H_2O, CO_2, S) are greatly enriched in the basaltic melts during intrusion and crystallization of the basalts. This is understandable considering the calculated amounts of fractionation. The volatile components contained in the differentiated residual melt reach the present outcrop level of the basalts. The points where these silicate melts were extruded onto the surface are only observed rarely (see also chapter 9). Thus, degassing of the melts was very restricted. The intrusions were nearly closed systems. The particular composition of the evaporites obviously enabled a portion of the volatile components in the magma to fix. Direct study is hence possible (chapter 6).

An attempt was made determining the extent of the theoretically possible enrichment of volatile phases in a cooling basaltic melt by assuming the following:
- A basaltic melt should be trapped in the upper beds of the salt body not far from the present outcrop level so that there is no chance for the volatile phases to degas at the surface (quasi-closed system).
- The dimensions of the basalt section to be used in the calculations are 0.5 m by 1 m by 30 km. The length of 30 km represents the extension of the dike through a fissure down to the crust/mantle boundary. Such a precondition likely yields values which are too low since the magma concerned originates from a depth of about 75 - 90 km (e.g., WEDEPOHL 1985).
- Fixing of the fluid phases in early crystallized minerals as well as the valency (oxidation number) of sulfur are disregarded.
- Further transport by water out of lower lying sediments is also disregarded.

The body of rock in question has a total volume of 15 000 m³ based on the measurements given above (1 m · 0.5 m · 30 000 m). As a basaltic melt has a density of about 2.5 g/cm³ (e.g., CERMAK & RYBACH 1982), the corresponding weight is 37 500 t. According to experimental studies and observations of natural basaltic glasses the initial portion of water in a basaltic melt probably amounts to about 1 % (e.g., SCARFE 1973; BREY & GREEN 1975; BURNHAM 1979; MUENOW et al 1979; BEYERS et al 1985; POREDA 1985). The portion of CO_2 that is also necessary for the formation of SiO_2-saturated melts is assumed to be 0.5 %. Consequently, the model basalt body could release a total of 380 t H_2O and 190 t CO_2. The initial amount of sulfur in basaltic melts could amount to about 800 µg S/g melt, i.e., about 30 t of sulfur for the model basalt.

When it is postulated that 0.1 wt% of the volatile phases reaches the present outcrop level due to the lacking possibility of degassing at the surface, 0.38 t H_2O, 0.19 t CO_2, and 0.3 t sulfur (the latter as sulfide, native sulfur, and sulfate) are released.

KNIPPING (1984) and KNIPPING & HERRMANN (1985) determined with the help of model calculations that during intrusion of a basaltic melt in a carnallitic rock a solution saturated with NaCl at 110 °C and about 200 kg of water can alter about twenty times as much salt rock. On this basis, the amount of salt rock that can be altered by the aqueous phase of the basalt calculated above can be estimated. It is assumed that a 2-m-thick potash salt seam (carnallitite) has a 1-m-long contact with the basalt. It follows that 1 m width of the basalt body · 2 m seam thickness · 2 (right and left contacts) = 4 m² of contact area. When 0.1 % of the aqueous phase (ca. 0.4 t) penetrates into the potash salt seam, 4 t of carnallitite can be altered. With a density of 2 g/cm³ this totals 2 m³. If this 2 m³ is distributed equally on both sides of the dike, the width of the contact zones amounts to 0.5 m. If 1 - 10 wt% of the volatile phases reaches the present outcrop level, the contact zone would be 5 - 50 m wide on both sides of the basalt dike.

This model calculation should give an impression of the dimensions of the partially extensive alterations of evaporite deposits through solution metamorphism and the enrichment of gases and native sulfur in evaporites from basaltic intrusions.

6 Isotope determinations

Great quantities of volatile phases (mainly H_2O and CO_2) have penetrated into the evaporites during intrusion of the basaltic melts. These phases have effected mineral reactions above all in the particulary reactive K-Mg rocks of the Hessen (K1H) and Thüringen (K1Th) potash seams of Zechstein 1. In doing so, partly great volumes of gases were trapped in the altered evaporites (e.g., Frantzen 1894; Kühn 1951; Hartwig 1954; Müller 1958; Herrmann 1979; Knipping & Herrmann 1985).

The enrichment of native sulfur in K1H and the overlying Begleitflöze in the area of dikes <D>, <N>, and <O> (fig. 2.2) is directly related to the intrusive events. The sulfur is found on the contact surfaces between the basalts and evaporites (see fig. 3.4). It also occurs up to 40m away from the basalt as fracture fillings or on the bedding planes in the evaporites. Käding (1962) already mentioned a presently inaccessible sulfur-filled fracture near dike <C>. The metamorphic sylvinite in the Hessen potash seam is impregnated in places with finely disseminated sulfur. The rock occasionally consists nearly completely of sulfur, occupying volumes of up to cubic meters (fig. 3.5).

Isotope studies were conducted for obtaining initial, reliable information on the genesis and origin of the native sulfur. Enrichments of polyhalite, kieserite, and langbeinite found near the basalt dikes or at the basalt-evaporite contacts were also included in the investigations. Determining the $\delta^{34}S$ values should explain whether the sulfur components of these minerals originated from the evaporite deposits, or whether they were introduced externally by the basalts. Part of the results discussed here have already been published (Knipping 1986).

It is noteworthy that the native sulfur always occurs together with gas-rich salt (*Knistersalz*). The mixtures of gas consist of CO_2, larger quantities of N_2, and subordinate hydrocarbons like CH_4 (Ackermann et al 1964). They are fixed in the evaporites in the following ways (Herrmann 1988):
1. *Fracture-bound* gases in fractures, fissures, and joints were already mentioned by Scheerer (1911) and Beck (1912b).
2. *Mineral-bound* gases. (i) Intercrystalline (intergranular), i.e., fixed between the grain boundaries of the minerals (Hartwig 1954), and (ii) incrystalline (ingranular, intragranular) in the form of microscopic bubbles in the mineral grain (Hartwig 1954, Oelsner 1961).

In contrast to the gas mixtures fixed in the evaporites, the solutions involved in the recrystallisation appear for the most part to have migrated out of the salt formation (e.g., Baar 1958; Herrmann 1979). On the other hand, these solutions are possibly dispersed over larger portions of the deposit in the form of microcrystalline inclusions as well.

As observed in several thin sections, the gas and CO_2 in samples used for $\delta^{13}C$ determinations in this study are fixed mostly in the incrystalline form. The isotope determinations should provide evidence of the source of the CO_2.

6.1 Sulfur isotopes

Native sulfur was observed underground at the Hattorf mine in K1H and in the overlying Begleitflöze at the contacts to basanite dikes <D> and <N> and to the phonolitic tephrite of dike <O>. Sulfur occurs only sporadically around dike <D>, but is abundant with up to 10^3 kg in the vicinity of dike <O>. The sulfur occurring together with sulfate minerals in several evaporite samples from around dikes <N> and <O> was investigated under the microscope.

It was found that the native sulfur is rarely and only initially hypidiomorphic. It occurs mostly as xenomorphic fillings between phenocrysts in the predominantly halitic groundmass. It appears as though a mist of sulfur invaded the evaporite matrix and then crystallized (fig. 6.1). Distinct grain boundaries of the halite are only rarely seen. As anticipated, swarms of oriented, elongate fluid and round gas inclusions (< 0.005 mm Ø) are very abundant in the halite.

In the potash seam K1H the evaporites, which have obviously been converted into halite, contain in addition to sulfur the sulfate minerals anhydrite and kieserite as well as subordinate langbeinite, polyhalite, and kainite. Sulfur is frequently intimately intergrown with the mostly hypidiomorphic sulfate minerals. In outcrop 60/147, the sulfur contains inclusions of idiomorphic anhydrite needles (fig. 6.2). Directly at the contact to the basalt, there are cm-wide zones composed nearly completely of anhydrite or finely acicular polyhalite with native sulfur, with chloride minerals (halite, sylvine) making up only a few percent.

The primary sulfate minerals (i.e. present before the intrusive event) cannot be distinguished clearly from the neogenic minerals without thorough mineralogical and chemical study. Only the kieserite of the Flockensalz in K1H near outcrop 60-ug (with strong effects of resorption) was able to be assigned to the original constituents. The mineral reactions caused in the potash seams by the basalt intrusions were interpreted by KNIPPING & HERRMANN (1985) and GUTSCHE & HERRMANN (1988).

Table 6.1 show the result of the sulfur isotope determinations. Several unpublished determinations on native sulfur and sulfates from the Kali und Salz AG works (Sigmundshall and Hansa) in the Hanover mining district, which were made available by DR. H. NIELSEN, are given for comparison.

In all sulfur-bearing evaporite samples the $\delta^{34}S$ values of the sulfates were also determined for detecting a possible reduction of the sulfate sulfur and thus an enrichment of ^{34}S in the resulting sulfate. Sample 23/51, in which the sulfate content was too low, is an exception. Three sulfate-bearing evaporite samples (anhydrite, kieserite), which contained no native sulfur, are used for comparison. Sample 62/146 was collected near a limburgite outcrop which was described by GUTSCHE (1988) and GUTSCHE & HERRMANN (1988). Samples 8/113 and 8/133 were taken from the first Begleitflöz of K1H near dike <N>.

Nearly all sulfur-bearing evaporite samples contained small amounts of pyrite which were not detectable in thin sections. Amounts of pyrite sufficient for sulfur isotope determination were only able to be isolated from sample 9/19. The pyrite here

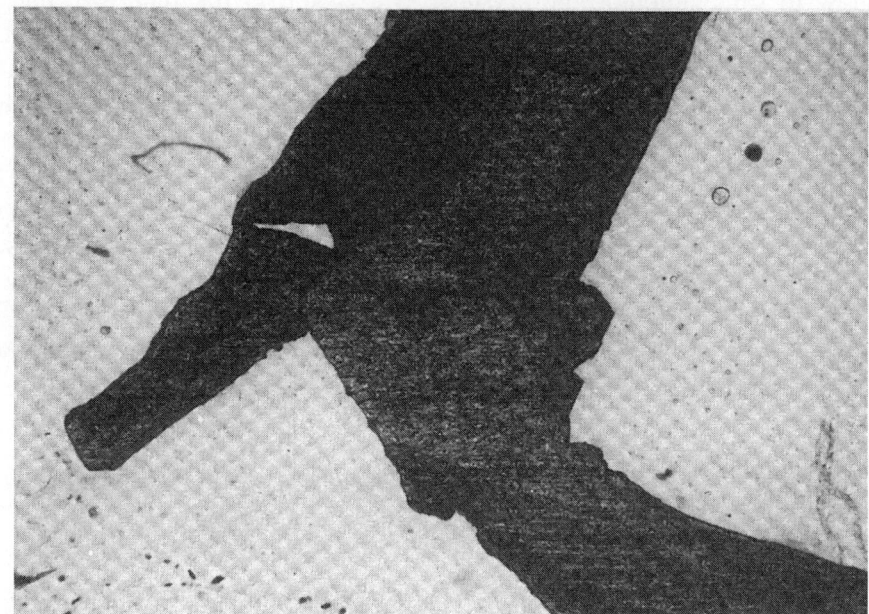

0,2 mm

Fig. 6.1 Outcrop 38-ug in K1H about 30 m away from phonolitic tephrite dike <O>. Native sulfur (dark) as filling between grains of halite (unpolarized).

0,2 mm

Fig. 6.2 Outcrop 60-ug in K1H. Native sulfur contains inclusions of fine needles of anhydrite.

grew in small idiomorphic crystals up to 1 mm Ø together with native sulfur and anhydrite along the contact to the hornblende-bearing basanite of dike <N>.

In some outcrops the basalt contained macroscopic deposits of native sulfur. An attempt was made to isolate this native sulfur from the basalt matrix. However, this was successful only in the case of basanite sample 23/52. Extractions from other samples yielded 2 - 10 mg of sulfur-yellow bituminous substance (also referred to as bitumen in the following) from 200 g of basalt. The $\delta^{34}S$ value was not able to be determined due to the low amounts of SO_2 and high amounts of extraneous gas following preparation for mass spectrometer measurements.

6.2 Carbon isotopes

The $\delta^{13}C$ determinations for CO_2 from three gas-bearing samples near dikes <N> (hornblende-bearing basanite) and <O> (phonolitic tephrite) together with the double-checks are given in table 6.2. In view of the simple preparation methods used, the agreement of the double-check is satisfactory. The $\delta^{34}S$ values for the three samples are given in table 6.1.

Table 6.2 shows the result of the $\delta^{13}C$ determinations for the bitumen samples extracted from the basalts as well. A high amount of extraneous gas interfered considerably with these measurements, as was also the case with the sulfur isotopes. Yet, measurement was possible in two instances (contamination controls using the recorded mass spectra). Table 6.2 also contains a $\delta^{13}C$ value for the Mg-Ca-Fe carbonates obtained from the groundmass of the unclassified basalt of dike <I>.

Table 6.1 Sulfur isotope measurements.

Sample numbers are arranged in ascending order within the dike. δ values for native sulfur and sulfates from the following localities are given for comparison (NIELSEN, unpublished):

(a) Sigmundshall mine, boundary between the Staßfurt potash seam (K2) and Unterer Leineton (T3),
(b) Sigmundshall mine, level 420 m, Ronnenberg potash seam (K3Ro), deposit E,
(c) Hansa mine, Staßfurt potash seam (K2), lower part of deposit, Hartsalz with violet kainite.

Explanations for the columns

1. <D>, dike enumeration (see also fig. 2.2); *BA*, basanite; *LI*, limburgite; *PT*, phonolitic tephrite.

2. Sample 9/19 is pyrite from the basalt-evaporite contact; (c) is H_2S.

5. Qualitative comparison of the portions of native sulfur in the total sample:

   + finest sulfur impregnations in the evaporites, ++ local enrichments of sulfur in the evaporites in centimeter to decimeter range, +++ nearly complete replacement of evaporites by sulfur in meter to decometer range. The mineral names designate samples in which the respective sulfate mineral takes on an unusually compact, rock-forming nature. When no mineral name is given, the sulfates are predominantly anhydrite and kieserite. *Basalt* designates native sulfur from the groundmass of the basalt.

   n.d. no data

	1	2	3	4	5
		$\delta^{34}S$ [per mil]			
	dike/ rock type	sulfide sulfur	native sulfur	sulfate sulfur	portion of S in total sample
15/34	<D>/BA		-2.8	+10.2	++
16/81	<J>/LI			+9.8	kieserite
16/82				+10.4	kieserite
62/146				+10.3	
7/115	<M>/BA			+11.5	langbeinite
8/16	<N>/BA		-3.2	+12.1	++
8/18			-2.8	+10.9	++
8/88			-2.6	+10.6	++
8/113				+10.1	
8/114			-2.7	+13.8	+
8/130			-6.2	+12.4	+
8/133				+10.1	
9/19		-3.1	-3.1	+11.5	++
13/28				+10.4	polyhalite
23/51			-3.3		+++
23/52			-3.3		basalt
23/53			-3.1	+11.3	+
24/54			-3.1	+11.2	++
28/62			-3.0	+13.1	++
38/831	<O>/PT		-2.1	+11.9	++
38/832			-1.4	+11.4	++
39/84			-7.1	+10.7	++
59/144			-2.0	+10.3	+++
60/142			-1.6	+17.4	+++
60/145			-9.3	+10.1	+++
61/143			-1.3	+12.7	+++
(a)	-		+0.3 +0.4	+11.5	n.d.
(b)	-		+1.0	+10.9	n.d.
(c)	-	-31.4	-27.8 -2.8	+9.6	n.d.

Table 6.2 Carbon isotope measurements.

Explanations for the columns

1. $<E>$, dike enumeration (see also fig. 2.2); *BA*, basanite; *ON*, olivine nephelinite; *PT*, phonolitic tephrite; *nc*, not classified basaltic rock. * Supplement from KNIPPING (1984).

2. The amounts of CO_2 used for each double-check of a sample were taken from the same container at intervals of 2 - 6 days.

4. The samples consisted of bitumenous, slightly sulfur bearing, undefinable substances (reduced carbon) which were extracted from the groundmass of the basalts with the Soxhlet method.

	1	2	3	4	5
		$\delta^{34}S$ [per mil]			
	dike/ rock type	CO_2 double- check	CO_2 mean value	bitumen from the basalt	Mg-Ca-Fe carbonates from the basalt matrix
10/22	$<E>$/ON			-28.6	
2D/24*	$<I>$/nc				-5.0
8/114	$<N>$/BA	-5.1/-6.0	-5.6		
8/17				-26.0	
60/142	$<O>$/PT	-7.2/-7.0	-7.1		
60/143		-8.4/-7.0	-7.7		
mean values			-6.8	-27.3	

6.3 Discussion

Since the C-isotope distributions provide suplementary evidence for interpreting the δ values for native sulfur, the $\delta^{13}C$ data will be discussed here first.

It has been known since mining in the Werra-Fulda district began (1901, shaft Kaiseroda I near Hämbach) that the Zechstein evaporites contain great amounts of gas (e.g., E. NAUMANN 1911, 1914; SCHEERER 1911; BECK 1912; GROPP 1919). Nevertheless, previous studies have been limited to a mostly qualitative description. ACKERMANN et al (1964) provided more recent compositional analyses of the predominantly CO_2 gas inclusions in the salt rocks. The following $\delta^{13}C$ values have been determined for CO_2: -17 per mil (MAAS 1962), -15 to -19 per mil (HOEFS 1973), and -6 to -19 per mil (KOCH & VOGEL 1980).

The fact that these results deviate greatly from the results of this study (about -7 per mil, table 6.2) can probably be attributed for the most part to a kinetic fractionation during sample preparation (see chapter 4.3). Furthermore, KOCH & VOGEL (1980; cf. MAAS 1962) found that samples with low values were contaminated with admixtures of hydrocarbons. According to ACKERMANN et al (1964) gas samples from the carnallitite and sylvinite of K1Th in the Werra district (Marx-Engels mine), for example, contained between 0.1 vol% and about 10 vol% CH_4 with a mean of 1 vol%.

Marine carbonates generally have $\delta^{13}C$ values of about 0 per mil (e.g., HOEFS 1980). Even greater values of about +3 to +7 per mil were determined in the Zechstein carbonates (e.g., MAGARITZ & SCHULZE 1980; MAGARITZ et al 1981). The CO_2 studied here could not have been mobilized out of the underlying, carbonate-bearing beds during volcanism because CO_2 produced during temperature-dependet reactions between silicates and carbonate is even heavier than the initial carbonate (SHIEH & TAYLOR 1969).

In carbonatites and kimberlites the $\delta^{13}C$ values usually lie between -4 and -7 per mil. These values are representative for unfractionated carbon from the earth's mantle. HOEFS (1973) obtained similar results for gaseous CO_2 from the Eifel. Since the values obtained in this study for the CO_2 inclusions from the evaporites also amount to about -7 per mil, this is an indication of magmatic origin.

The Ca-Mg-Fe carbonates obtained from the groundmass of the unclassified basalt from dike <I> are products of the decomposition of silicate minerals. The $\delta^{13}C$ value of -5 per mil (table 6.2; KNIPPING 1984) shows that the CO_2 involved in the formation of the carbonates originated from the silicate melt.

As seen in table 6.2, the reduced carbon of the bitumens has a $\delta^{13}C$ value of about -27 per mil. For volcanic methane HULSTON & MCCABE (1962), for example, found an average isotope composition of -25 per mil. For reduced carbon in about 50 different magmatic rocks HOEFS (1973) obtained comparable values of -25 to -29 per mil. Here, the similarity between these values and those for reduced carbon in carbonaceous chondrites and lunar rocks is pointed out. The occurrence of reduced, 'organically bound' carbon in basalts can be explained in the following way according to HOEFS (1973): This involves either (i) organic components mobilized by the silicate melts (the $\delta^{13}C$ for C_{org}, for example, in Kupferschiefer average -27.5 per mil according to MAROWSKY 1969), or (ii) the inorganic formation of juvenile carbon. However, the analytical values for continental basalts obtained with the recently developed *stepwise combustion method* (HOEFS, personal communication, 1987) are not yet available to aid in a clear explanation.

The results of the S-isotope determinations in table 6.1 are depicted graphically in fig. 6.3. The isotope distribution shows two independent maxima:

1. The $\delta^{34}S$ distribution of all sulfate samples has a frequency maximum between +10 and +13 per mil. Only sample 60/142 with +17.4 per mil deviated from the mean distribution.
2. The native sulfur from the evaporites and the basalt as well as the pyrite from the basalt contact have frequency maxima at -3 to -4 per mil. Three samples deviate greatly from this toward more negative values. Sample 8/130 contained only traces of native sulfur in addition to anhydrite and dominant halite. Sample 60/145 is composed of sulfur intergrown with compact langbeinite.

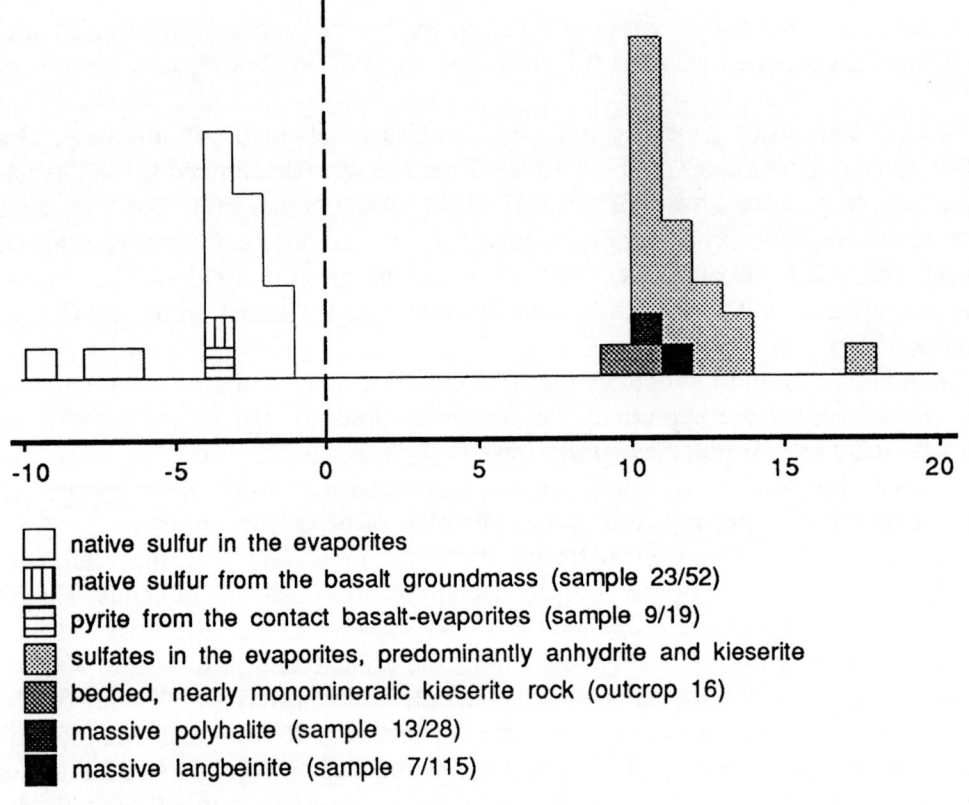

native sulfur in the evaporites
native sulfur from the basalt groundmass (sample 23/52)
pyrite from the contact basalt-evaporites (sample 9/19)
sulfates in the evaporites, predominantly anhydrite and kieserite
bedded, nearly monomineralic kieserite rock (outcrop 16)
massive polyhalite (sample 13/28)
massive langbeinite (sample 7/115)

Fig. 6.3 Sulfur-isotope distribution from native sulfur, pyrite, and sulfates from the Werra sequence in the Hattorf mine ($\delta^{34}S$ values in per mil).

The isotopic composition of the Zechstein sulfates, which are depleted in $\delta^{34}S$ compared with marine sulfates from other geological formations, is well known from the studies of NIELSEN & RICKE (1964), NIELSEN (1965), and PUCHELT & NIELSEN (1967). According to these authors the $\delta^{34}S$ values of the sulfate at the level of the potash seam amounts to about +8 to +13 per mil with a maximum between +11 and +12 per mil. This appears to be a clear indication of the fact that the sulfate components in the samples analyzed here originated from the deposit.

Since recystallization of the sulfate minerals does not effect a fractionation of sulfur isotopes, nothing can be said about the age of formation. The occurrence of the studied sulfate minerals, which form in part monomineralic rocks and the xenomorphic form of the crystals, are indicative of neogenesis or recrystallization. Hence, the other components in the sulfate minerals, e.g., calcium, can have originated from the decomposition of the silicate minerals of the basalts. The development of isothermal and polythermal reaction models in conjunction with chemical analyses are necessary for obtaining further information on the genesis especially of massive sulfates kieserite, langbeinite, and polyhalite and of the accumulations of anhydrite at the basalt contact. This is planned in a subsequent study.

In connection with the findings discussed in chapter 5.3 and the distribution pattern of the sulfur isotopes from the Hattorf mine (fig. 6.3) the following important statements regarding the genetic interpretation of the native sulfur can be made:

- With few exceptions the S-isotope distribution for all outcrops near the basalt dikes is uniform.
- As the magnetic measurements of SIEMENS (1971) and new absolute age determinations (chapter 7) have shown, the basaltic magmas have intruded the Zechstein evaporites at various geological times. Nevertheless, the measured samples have a uniform $\delta^{34}S$ distribution.
- Sometimes abundant sulfur occurs with limited sulfate while at other localities this relationship is reversed. Yet, the $\delta^{34}S$ values only vary slightly.
- Native sulfur always occurs together with gas-rich evaporites (*Knistersalz*).
- As the C-isotope distribution shows, the CO_2 from the gas inclusions in the evaporites can be derived from the fluid phases associated with the basaltic magmas.
- The greatest amounts of sulfur are found at the contact between the evaporites and the phonolitic thephrite (dike <O>). The phonolitic tephrite is the most fractionated rock (magmatic differentiation) at the Hattorf mine (cf. chapter 5.3). Through fractionation the volatile constituents in the magma may become highly concentrated in the resulting magma.

Three genetic models for the native sulfur will be discussed in the following:
1. bacterial reduction of sulfate,
2. inorganic reduction of sulfate,
3. derivation from the mobile phases of the basaltic magmas.

Bacterial reduction of sulfate

The commercially most important sulfur deposites are bound to occurrences of marine evaporits (FIELD 1972). In most cases the native sulfur occurs together with gypsum and calcite in the caprock of salt domes. The sulfur forms by the reduction of the sulfates (anhydrite, gypsum) to H_2S by anaerobic bacteria (e.g., FEELY & KULP 1957; SCHNEIDER & NIELSEN 1965). The principle of the bacterial reduction of sulfate has been discussed in detail, for example, by NIELSEN (1981, 1985). The fractionation occurring during this process can amount to as much as -60 per mil. The neogenic H_2S reacts with the other sulfate to form native sulfur (FEELY & KULP 1957). The fractionation factor during this oxidation is negligibly small. Organic substances (e.g., oil) are necessary as nourishment for the existence of the bacteria. The bacteria obtain the energy necessary for metabolism from the reduction of the sulfates. The CO_2 produced by oxidizing organic substances together with the calcium of the sulfates forms calcite. However, the most important feature of the bacterial reduction of sulfates is the greatly variable $\delta^{34}S$ values for the differing sulfates or sulfur crystals, which are observed in part within the closest confines, e.g., a hand specimen. Thus, such processes are conceivable for the genesis of the samples used for comparison in table 6.1 (NIELSEN, unpublished). Contrarily, the bacterial reduc-

tion of sulfate can be disregarded as a genetic model for the native sulfur from the Hattorf mine for the following reasons:
- The nutritional basis for bacterial activity which could have produced the great amounts of sulfur was lacking in the evaporites.
- CO_2 produced by bacterial metabolism generally has highly variable and/or low $\delta^{13}C$ values (-25 per mil on the average, e.g., HOEFS 1987, p. 36). On the other hand, about -7 per mil was measured for the CO_2 trapped in the salt (table 6.2).
- As indicated in fig. 3.4, native sulfur is definitely related to the intrusive events and hence formed in the evaporites at various geological times. Nonetheless, it has uniform $\delta^{34}S$ values.

Inorganic reduction of sulfate
In addition to a reducing agent the inorganic reduction of sulfate requires temperatures of at least 250 °C due to the high activation energy of the S=O bond (e.g., HOEFS 1987). The amount of fractination in an open system is about 15 per mil (e.g., NIELSEN 1981). This value corresponds approximately to the difference in $\delta^{34}S$ values between the studied native sulfur and sulfates.

The native sulfur cannot have formed at the place where it is exposed today. Simple considerations of mass balances show that the amount of initial sulfate in situ in the evaporites was too small to form the partly large amounts of sulfur ($> 10^3$ kg) intergrown with minor sulfate. Furthermore, there is no suitable reducing agent in the evaporites. Consequently, providing that the inorganic reduction of sulfate was the dominant process during formation, such a reaction must have occurred in the underlying strata (e.g., Unterer Werra Anhydrit, A1). The Fe(II) in the basaltic melts, for example, could have served as a reducing agent. Neogenic H_2S formed by reduction would subsequently be reoxidized producing native sulfur. The temperatures necessary for the inorganic reduction of sulfur were surely reached at the contact between the silicate melt and host rock. Yet, according to the calculations in chapter 8.4 it is improbable that the heat of the rapidly cooling, partially chilled basaltic melts was sufficient for reducing the sulfates to a great extent and then transporting the generated native sulfur to the present outcrop level (up to 40 m away from the basalt contact). The inorganic reduction of sulfate (seawater) is observed today in mid-oceanic-ridge basalts (e.g., ZIERENBERG et al 1984). Hence the formation of native sulfur by inorganic reduction of sulfate cannot be completely ruled out at least as a locally effective process.

Derivation from the mobile phases of basaltic melts
The S-isotope distribution of basalts as well as the chemical composition and $\delta^{34}S$ determinations of volcanic gases have been reported, for example, by SAKAI & NAGASAWA (1958), SCHNEIDER (1970), NORDLIE (1971), SAKAI et al (1982), FAURE et al (1984), SAKAI et al (1984) and HARMON et al (1987). However, these studies deal nearly exclusively with the sulfide and sulfate sulfur in basalt or the H_2S and SO_2 in volcanic gases. Native sulfur is not found in subrecent basalts because of its high susceptibility to weathering. Studies of recent volcanic gases show that SO_2 occurs

as a main sulfur-bearing phase besides subordinate native sulfur and minor H_2S. However, native sulfur is frequently sublimated already before or just after the gases exit at the surface. NORDLIE (1972) points out that native sulfur frequently constitutes a much greater portion of the total sulfur contents than is measured in the volcanic gases at the surface.

According to UEDA et al (1979) the $SO_2 - S_2 - H_2S$ balance in the volcanic gases is dependent upon temperature, pressure, oxygen fugacity, and the amount of water in the melt. For example, the amount of S_2 increases with the decreasing temperature of the volatile phases. However, UEDA et al (1979) do point out that the amounts of fractionation between the various sulfur phases are unknown or can only be estimated very roughly.

In the solfataras of Greece HUBBERTEN et al (1975) obtained $\delta^{34}S$ values for native sulfur of 0 to +8 per mil. Based on further observations they concluded these high values can be attributed to contamination by seawater sulfate (+20 per mil). SAKAI et al (1982) determined that native sulfur from the Kilauea volcano of Hawaii has $\delta^{34}S$ values of -7 to -13 per mil. UEDA et al (1979) ascertained a mean $\delta^{34}S$ value of -4 per mil for 140 samples from Japanese volcanoes. The samples from the Hattorf mine exhibit a similar isotope distribution (table 6.1, fig. 6.3). A value of about +1 per mil is predicted for sulfur in the earth's mantle (e.g., SCHNEIDER 1970; GRINENKO et al 1975). Assuming that the native sulfur was derived from the mantle, it has been fractionated by about -5 per mil compared with mantle sulfur. A clear explanation for this fractionation which may be typical for the formation of native sulfur has not yet been found (UEDA et al 1979).

The $\delta^{34}S$ values for the pyrite from the basalt contact (-3.1 per mil) is comparable with the data for native sulfur. The minutest amounts of pyrite, which were, however, insufficient for isotope determinations, were found in practically all sulfur-bearing evaporites. H_2S in the silicate melts has possibly been fixed by small amounts of iron during the alteration of the evaporites. As the minute amounts of pyrite show the amounts of H_2S in the mobile phases were insignificant at the time of intrusion. The amounts of H_2S in the gas inclusions of the evaporites of the Werra region are less than 0.1 vol% (ACKERMANN et al 1964).

In this context, the question of what happens to the SO_2 in the silicate melt must finally be asked. Since SO_2 is generally the dominant sulfur phase in uncontaminated volcanic gases, the $\delta^{34}S$ value lie typically between 0 and +1 per mil (e.g., SAKAI et al 1982). In the case of the basaltic melts in the Werra region the SO_2 of the mobile phases could have been fixed as sulfate in the subsurface. SCHNEIDER (1970) determined the $\delta^{34}S$ value of the sulfate sulfur in the basalts of the Northern Hessian Depression to be about 0 to +5 per mil. However, the values in table 6.1 for the sulfates (anhydrite, kieserite) intergrown with native sulfur are precisely typical of those for Zechstein sulfate. Either the $H_2S - S_2 - SO_2$ equilibrium shifted extremely in the direction of native sulfur, or large amounts of SO_2 have served as an oxidation agent for constituents in the silicate melts (see, e.g., NORDLIE 1971). After the SO_2 has been reduced to native sulfur, it could then find its way into the evaporites with the volatile phases.

Following studies with the microscope the native sulfur appears to have been injected into the salt rocks. It was possibly melted in a high-temperature aqueous phase and migrated into the evaporites during solution processes. A similar principle is applied when recovering sulfur with the so-called *Frasch process*, as is well known. The occurrence of native sulfur up to 40 m away from the basalt contact also shows that it can only have been transported under high pressure and thus high velocity. During the slow migration of sulfur-bearing aqueous phases the native sulfur would already have cooled and crystallized near the basalt contact due to the difference in temperature between the basalt magma and host rock (chapter 8). The high pressure which must have existed during intrusion of the silicate melts is also evidenced by other observations, like the occurrence of relictic carnallite at the basalt contact (KNIPPING 1984; KNIPPING & HERRMANN 1985) or the occurrence of tuffs (e.g. outcrop 44-sf, Trumbachsköpfchen). Regarding the question of pressure it is perhaps interesting that during underground atomic bomb tests in shallowly dipping evaporite beds (Gnome project south of Carlsbad, New Mexico; RAWSON 1963) and in a salt dome (Salmon experiment, Lamar County, Mississippi; RAWSON et al 1966, 1967) molten NaCl was also pressed into the fissures and cracks exactly 20 - 60 m into the surrounding salt rocks.

In summary, the sulfur and carbon isotope studies have provided evidence that the native sulfur in the Hattorf mine is 'juvenile' sulfur, which invaded the evaporites during intrusion. In the discussion of the results a first attempt at interpretation is made which can serve as the basis for the necessary supplemental studies (see chapter 10).

As the results discussed in this chapter have shown, especially the alteration processes of the K-Mg rocks of the potash seam were suitable for preserving the mobile phases of the basalt magmas. This can surely be attributed to the fact that the basaltic melts only had very limited possibilities for degassing since there was no link to the surface. As in the case of submarine basalts, further studies might reveal more about the mechanism of intrusion and the composition of the volatile phases in the basaltic melts.

7 K-Ar age determinations on the basaltic rocks

Magnetometer measurements taken on the basalt dikes at the surface in the area of the Hattorf mine showed that the dikes have differing magnetic signatures (e.g., SIEMENS 1971). Geomagnetic reversals occur on the average of about every 300 000 to 500 000 years. This is an indication that the intrusion of basaltic magmas into the evaporites of the Werra mining district was not an isolated event, but occurred at various geological times. This finding is supported by the knowledge that the mineralogical and chemical composition of the surface and subsurface basalts in part evidences magmatic differentiation of various degrees (chapter 5.3). However, for fractionation to occur the parent magmas must stagnate at different times during ascent from the mantle. Absolute age determinations on the basalts appear desirable to obtain more on the temperatures in the deposit (chapter 8) and on the intrusive mechanism of the silicate melts (chapter 9).

The whole-rock datings were done with the K-Ar method by Teledyne Isotopes (Westwood, N.J., USA). To avoid argon loss measurements were taken on rock fragments between > 1.25 mm and < 2 mm in size. According to LIPPOLT (personal communication, 1986) the subsurface basalts cannot be dated because they have absorbed Ar from the K-Mg rocks of the potash seams. According to the study presented here one cause for ages which are too high could be the incomplete degassing of the silicate melts and thus the fixing of argon from the mantle during solidification. According to the studies of BROOKINS (1981), on the contrary, the values measured during K-Ar age determinations on Tertiary basalts ('lamprophyres') in Permian evaporites (Salado Formation near Carlsbad, New Mexico) were not falsified. Consequently, two subsurface and five surface samples were selected for the determinations on the basalts from the Werra mining district. The results are given in table 7.1.

The dated subsurface outcrops 53-ug and 8-ug are situated nearly directly below surface outcrops 43-sf and 44-sf (see fig. 2.2). However, the subsurface samples yielded an age of 44.2 Ma and 35.7 Ma, whereas the surface samples from each dike were dated at 24.9 Ma and 17.5 Ma. Thus, either the subsurface samples absorbed argon from the host rocks, or the silicate melts did not release their primary argon completely following intrusion.

An extensive discussion of Miocene volcanism in the Northern Hessian Depression in light of age datings including data on the reliability of the values is given by WEDEPOHL (1982). He concluded that the tholeiitic basalts with an age of 10 - 20 Ma are the oldest dikes in this region. The peak of volcanic activity 11 - 15 Ma ago involved alkali-olivine basalts. The olivine nephelinites, limburgites, and basanites with ages of 7 - 14 Ma are the youngest basalts of the Northern Hessian Depression. The volcanism in the Vogelsberg region (see fig. 2.1) is of similar age. The Rhön (see fig. 2.1) volcanism, in contrast, already begun 25 Ma ago, reached its peak between 22 - 19 Ma, and lasted sporadically until 11 Ma (LIPPOLT 1978, 1982). The produc-

Table 7.1 Whole-rock ages of seven basalt samples from the Werra-Fulda mining district.

ug subsurface (underground) exposure
sf surface exposure
<L> dike enumeration (cf. fig. 2.2)
BA basanite
ON olivine nephelinite
PT phonolitic tephrite
nc not classified

The whole-rock age ($\bar{x}$) is given with the absolute error of the analytical procedure (± s)

exposure/ dike	basalt type	sampling site	whole-rock age [Ma], $\bar{x}$ ± s
43-sf/<H> 53-ug/<H>	BA	SW Hohenroda farm near Oberbreitzbach, 366 m a.s.l. Südfeld entrance via 6th south tunnel, below surface outcrop 43-sf	24.9 ± 1.3 44.2 ± 2.2
42-sf/<L>	ON	Hohenroda-Soisliden, SE slope of Soisberg	15.5 ± 0.8
44-sf/<N> 8-ug/<N>	BA	road from Ransbach to Schenklengsfeld, junction Wehrshausen, Trumbachsköpfchen district 4W entrance, G47, below surface outcrop 44-sf	17.5 ± 0.9 35.7 ± 1.8
40-sf/<O>	PT	NE Oberlengsfeld, at the eastern wall of Landeck castle	22.5 ± 2.0
46-sf/<P>	nc	sample 46/86 (potential alkali-olivine basalt), SSW Friedewald, Dreienberg, abandoned quarry behind rifle club house	13.6 ± 0.7

tion of various basaltic melts at different times can be attributed to a tectonic stress field active in the mantle at greater and greater depths. Thus, the various degrees of dissolution of the mantle rock led to the formation of silicate melts of differing composition, i.e., different types of basalt. The tholeiitic basalts studied by WEDE-POHL (1982, 1983, 1985, 1987) formed by differentiation of an olivine-tholeiitic melt. The formation of the latter requires that about 13 % of the mantle rock is dissolved. The depth of formation of olivine-tholeiitic melts is assumed to be 30 - 45 km. The youngest basalts require the lowest degree of dissolution, i.e., 4 - 6 %, and form at depths of 75 - 90 km.

The zones with the older tholeiitic basalts and alkali-olivine basalts in the Northern Hessian Depression west of the Hattorf mine strike NNE - SSW. In contrast, the zones with the younger SiO_2-undersaturated basalts (olivine nephelinites, limburgites, basanites) strike NNW - SSE (WEDEPOHL 1982). The surface samples dated for this study are similar mineralogically and chemically to the SiO_2-undersaturated basalts used for comparison. However, they are older, like the tholeiitic basalts used for comparison. As seen in fig. 2.2, the dominant strike of the basalt dikes at the Hattorf mine is comparable with the strike of the older basalts further west.

Consequently, it is clear that in the Werra region a tectonic stress field already extended down to greater depths in the mantle, even before the production of the alkali-olivine-basaltic melts further to the west began. As SiO_2-undersaturated basaltic melts then ascended from the mantle to the earth's surface about 14 Ma ago also west of the Northern Hessian Depression, the stress field effective at depths of 75 -90 km changed its strike to NNW - SSE. The production of basanitic and olivine-nephelinitic melts, which formed Trumbachsköpfchen and Soisberg had already ceased by this time. Regarding age, the basalts of the study area would obviously have to be classified with the Rhön volcanites.

Since the basalts studied here, particularly the potential alkali-olivine basalt (46/86), resulted mostly from fractional magmatic differentiation of differing degree, the ages determined only reflect the sequence of intrusion and not necessarily the sequence of melt formation. Yet, the results presented here are by all means an indication of the temporally delayed formation of the basalts in the individual dikes in the Werra mining district. This means that the host deposits suffered the effects of the intrusive events at different geological times.

8 Model calculations of the spatial and temporal temperature distribution at the basalt-rock salt contact after intrusion of the basaltic melts

Effects of volcanic events can be seen in various occurrences of evaporites. Models for calculating mass balances in evaporite bodies can be inferred from the related alteration and redistribution processes, above all in the salt rocks (KNIPPING 1984; KNIPPING & HERRMANN 1985; HERRMANN 1987; GUTSCHE & HERRMANN 1988; KNIPPING 1989). Hereby, the reaction temperatures are of fundamental importance. As yet, there have only been assumptions and estimates of the temperatures which were present in the basalt-evaporite contact zone. For this reason an attempt was made at quantifying the spatial and temporal temperature distribution after intrusion of the basaltic melts with model calculations.

JEAN BAPTISTE JOSEPH FOURIER (1768-1830) had already discussed the equation for thermal conductivity in his 'Théorie analytique de la chaleur' (1822). This strictly mathematical procedure was employed by LANE (1899) for calculating the temperatures of cooling sills. In 1913 INGERSOLL & ZOBEL published the first simplified and geologically applicable solutions to the 'Fourier equation' for two-dimensional intrusive bodies.

Subsequently, other authors in the geosciences were occupied with heat propagation (e.g., LOVERING 1935, 1936, 1955; WINKLER 1949a, b, c, 1967; JAEGER 1957, 1959, 1961, 1964, 1968; MUNDRY 1968; JOBMANN 1985).

Comprehensive and geologically significant papers on the conduction of heat in solid bodies were published by INGERSOLL et al (1954) and CARSLAW & JAEGER (1959). Other works ensued from questions concerning the storage of heat-producing radioactive wastes in rock salt (e.g., EDWARDS 1966; SCHMIDT 1971; DELISLE 1980). An introductory overview of 'geothermy' (in German language) is given by BUNTEBARTH (1980).

Many of the aforementioned authors employed analytical methods in their calculations. However, CARSLAW & JAEGER (1959, p. 466), SCHMIDT (1971), and BUNTEBARTH (1980) point out that only simple problems concerning thermal conduction can be solved analytically. Solutions to complicated problems with numerical methods (as defined by SMITH 1965, 1970), like those applied by, for example, SCHMIDT (1971) and DELISLE (1980), are limited by the capacity and above all the calculating speed of the available computer. Some problems were first able to be solved recently with the improvements in computer technology. A numerical procedure was also chosen for the calculations in this study to be able to take the temperature dependency of the thermal conductivity of halite into account.

8.1 Assumptions

Even a complex mathematical model is not able to allow for all the interactions of the physical and chemical processes occurring during intrusion of a magma. Since, in addition, several parameters are inaccurately or not at all known (see chapter 8.5), a compromise between sufficient accuracy and mathematical expenditure must be found using simplified assumptions. They are the following:

1. At the Hattorf mine the basalt dikes generally strike in a nearly NS direction. Therefore, the dikes in the model are assumed to have an infinite vertical and NS extension (i.e., a plate of constant thickness and infinite extension).

2. The host rock is a homogenious rock salt (melting point of NaCl is 801 °C, e.g. D'Ans et al 1967). The anhydrite, clay minerals, and other accessory components in the rock salt of the Werra sequence (Na1) are disregarded. In the Werra region the Na1 is up to 300 m thick (e.g., Käding 1978).

   Assuming in a simplified way that the intruding basaltic melt was 'dry' (see point 6 below) the melting point of several evaporite minerals near the contact to the K-Mg rocks will be exceeded (data on release temperatures of crystal water and melting temperatures compiled in Herrmann 1983, p. 146). In this way the magma could actively provide for more space dependent on the temperature and mineralogical composition of the host rock and also without the participation of an aqueous phase. The extent of such processes has not yet been able to be treated in a mathematical model. Hence, no calculations were done for the comparatively thin potash seams (K1Th, 2 - 10 m; K1H, 2 - 3 m).

3. It is assumed that intrusion occurred rapidly and came immediately or very quickly to a stop (cf. Kühn 1951). The heating of the host rock by a fluid phase preceding the melt can be disregarded due to the great ascent velocity of the magma from the earth's mantle or from a reservoir at grater depth (magma chamber; cp. capters 5.3, 9). Comparatively few points of exit of the basalt magma have been found at the surface in the Werra region (see fig. 2.2). A great portion of the melts solidified in the strata below the present outcrop level. Consequently, the host rock was only locally heated over longer periods of time by the constant supply of silicate melts. Hence, this not considered in the model. Multiphase intrusions, i.e., the intrusion of magma into an already solidified rock of the same origin and composition following the formation of new paths (e.g., Käding 1962; Koch & Vogel 1980), were not able to be substantiated in the subsurface outcrops studied.

4. Convection as the magma cooled does not have to be considered in the calculations because the dikes in the subsurface outcrops have a maximum width of 1.8 m (outcrop 50-ug).

5. Even macroscopically relatively fresh samples have a completely vitrified zone in the cm range at the contact. Similar observations were also made for submarine magma extrusions, i.e., pillow lavas. Larger crystals (pyroxenes, hornblendes) are frequently enriched in the interior of the dikes. The groundmass of many samples, especially of the limburgites, consists completely of glass. During intrusion the

magma was obviously chilled due to a difference in temperature of more than 1 000 °C between the basaltic melt and the host rock (chapter 8.3). Differences in the thermal behavior between the glass phase and the magma or the crystalline silicate rock were neglected in the calculations.

6. 'Dry' basalts do not occur in nature. The primary amounts of water in frequent basaltic melts may be about 1 % (cf. chapter 5.3). Since only a few dikes outcrop at the surface in the Werra region and these rocks are frequently differentiated, mobile components in the magma (e.g., H_2O, CO_2) as well as possible aqueous solutions from the underlying strata near the solidified subsurface melts were able to be enriched. In many cases the spreading out of the basalts leading to sills, fluid and gas inclusions in the salt rocks, neogenic minerals, changes in the volume of the potash seams, and partly extensive impoverishments is attributed to the interactions between fluid aqueous phases and the evaporites at the time of active volcanism (e.g., KÜHN 1951; HARTWIG 1954; BAAR 1958; MÜLLER 1958; HOPPE 1960; D'ANS 1967; BRAITSCH et al 1964; HERRMANN 1979, 1980; KNIPPING & HERRMANN 1985).

The silicate melts must have cut through an average of about 150 m of rock salt beds (Na1α, Na1β) to reach the level of the present subsurface outcrops, which are mostly in K1H. It is not yet fully clear whether an accompanying fluid phase was able to be saturated with NaCl during ascent of the magma (KNIPPING 1984; KNIPPING & HERRMANN 1985; chapter 5.3). Mineral reactions and migration can be identified above all in the potash seams by comparing the mineralogical and chemical composition of the evaporites at the basalt contact with unaffected parts of the rock. With the help of solution equilibria of marine evaporites (D'ANS 1933), BRAITSCH (1962, 1971), for example, quantitatively worked out important metamorphic processes. Mass-balance calculations enabled KNIPPING (1984) and KNIPPING & HERRMANN (1985) to estimate the amount of solution which penetrated the Thüringen potash seam during basalt intrusion for a certain amount of rock. Such a solution dissipated heat not only through the pore space of the rock, but also through cracks and fissures (e.g., LOVERING 1955; JAEGER 1959). Estimating the amount of heat transported is also hindered by the solubility of the salt.

In this context the question of the solution velocity in the evaporites under the pressure and temperature conditions of a magma intrusion arises. As yet, only little experimental and calculated data are available (e.g., KARSTEN 1954; SDANOWSKI 1958; HOFFMANN & EMONS 1969; HOFRICHTER 1974; HENTSCHEL & KLEINITZ 1976; RÖHR 1980). These data concern, for example, aspects of constructing salt caverns for storing fluid and gaseous hydrocarbons and of compressed air and the subsurface solution mining of salt. The results are inapplicable to the questions raised in this work because they were obtained in an open system with unsaturated aqueous solutions and under low pressures.

A further complication is given by the question of whether aqueous solutions always intruded the evaporites at the same time as the magmas. It is conceivable, for example, that the solutions and gases ascended before or after the intrusion of the magmas.

It follows from the statements above that the effects of heat transport and distribution through fluid phases on the absolute temperature and spatial temperature development in the evaporites cannot be understood exactly. Although the processes described surely led to an accelerated cooling of the silicate melt and a decrease in the heating up of the evaporites (chapter 8.5), the aqueous phases were not able to be considered in the calculations. *Consequently, the model calculations primarily provide information on the maximum temperatures expected in the evaporites following intrusion of 'dry' basaltic melts* (cf. Knipping 1987, 1989).

8.2 Mathematical models

The FORTRAN computer program ICTF ('Iterative Calculation of Temperature Fields') was developed on the VAX computer at the Institute for Numerical and Applied Mathematics and on the PDP 11/34 at the Institute of Geochemistry of the University of Göttingen for carrying out the computations.

The basis of the calculations lies in solving partial differential equations with the help of a so-called finite difference approximation. The temperatures for defined points in the basalt body and host rock are calculated stepwise for a certain time interval with the known values according to equation (8.1) (modified from SMITH 1965, p. 10, 1970, p. 22). Figure 8.1 illustrates the principle of the calculations. For $T_l < T_s < T_n$ the following is true:

$$T_{new,s} = T_{old,s} + \frac{\Delta t}{\rho\, c\, \Delta s^2} \left((T_{old,s+1} - T_{old,s})\, \lambda - (T_{old,s} - T_{old,s-1})\, \lambda \right) \tag{8.1}$$

T	temperature in °C
$T_{new,s}$	actual temperature at point s calculated on the basis of T_{old}; this value differs from T_{old} by the elapsed time Δt and is thus used for calculating the subsequent T_{new}
$T_{old,s}$	known temperature at point s before the elapsed time Δt
t	time in seconds; t_0 = time of intrusion
Δt	time interval
ρ	density of basalt or host rock in kg/m^3 (= g/cm$^3 \cdot 1\,000$)
c	specific heat capacity of basalt or host rock in Ws/kgK (= kJ/kgK $\cdot 1\,000$)
s	distance from outer contact in m (see fig. 8.1)
Δs	distance interval
$T_{old,s+1}$, $T_{old,s-1}$	known temperatures at distance Δs to the right and left of point s at time Δt before the actual time
λ	thermal conductivity of the melt or host rock in W/mK

The temperature fields were calculated for one side of the basalt dike in each case. The temperature field is symmetrical with respect to the central plane of the basalt dike. Thus,

$$T_{new,1} = T_{new,K} \tag{8.2}$$

can be used (see fig. 8.1). Moreover,

$$T_{new,n} = T_{new,n-1} \tag{8.3}$$

is valid with sufficiently great distance from the contact (JOBMANN 1985).

In contrast to an analytical method (e.g., CARSLAW & JAEGER 1959, p. 54, eq. 3), the numerical method used here allows the temperature dependence on thermal conductivity of the rock salt to be considered.

Although the relatively wide dikes were preferred for sampling, about 40 % of them are only 20 - 40 cm wide. Only 10 % of the dikes are wider than 1 m (see also KOCH & VOGEL 1980, p. 73). Hence, the calculations for the rock salt were limited to a distance of 20 m from the basalt contact. However, due to the thinness of the dikes, small distance and time intervals had to be selected (chapter 8.3), since the method would otherwise become numerically unstable (SCHMIDT 1971). The numerical stability was tested by varying the distance and time intervals.

The great calculation time was caused by the small distance and time intervals. For example, for a 1-m-thick dike the calculation time on the PDP 11 was about 50 hours with the following selected parameters (see also chapter 8.4):
- calculated temperature field extended 20 m into the rock salt,
- the distance interval was 0.1 m,
- the time after intrusion was 1 year, and
- the time interval was 60 seconds.

In this example, 210 temperature values were calculated for each of about $5 \cdot 10^5$ time intervals. Calculation time for tests and short-time calculations were able to be reduced considerably by reducing the temperature field to be calculated. The following algorithm then served as an alternative to (8.3) for calculating $T_{new,n}$:

$$d_1 = T_{new,n-3} - T_{new,n-2}$$
$$d_2 = T_{new,n-2} - T_{new,n-1}$$
$$p_1 = d_2 \cdot 100/d_1 \tag{8.4}$$
$$p_2 = p_1 \cdot d_2/100$$
$$T_{new,n} = T_{new,n-1} - p_2$$

Thus, the difference in percent between the differences, d_1 and d_2 is also calculated for the difference between $T_{new,n-1}$ and $T_{new,n}$. However, this algorithm cannot be used for the time after intrusion as the thermal conductivity of the host rock (rock salt) at the contact is lower than that of the basalt (contact temperature > 400 °C). This type of calculation would hence yield incorrect results. If the temperature at the contact is less than 400 °C, the temperature field to be calculated must be enlarged to be valid for equation (8.3).

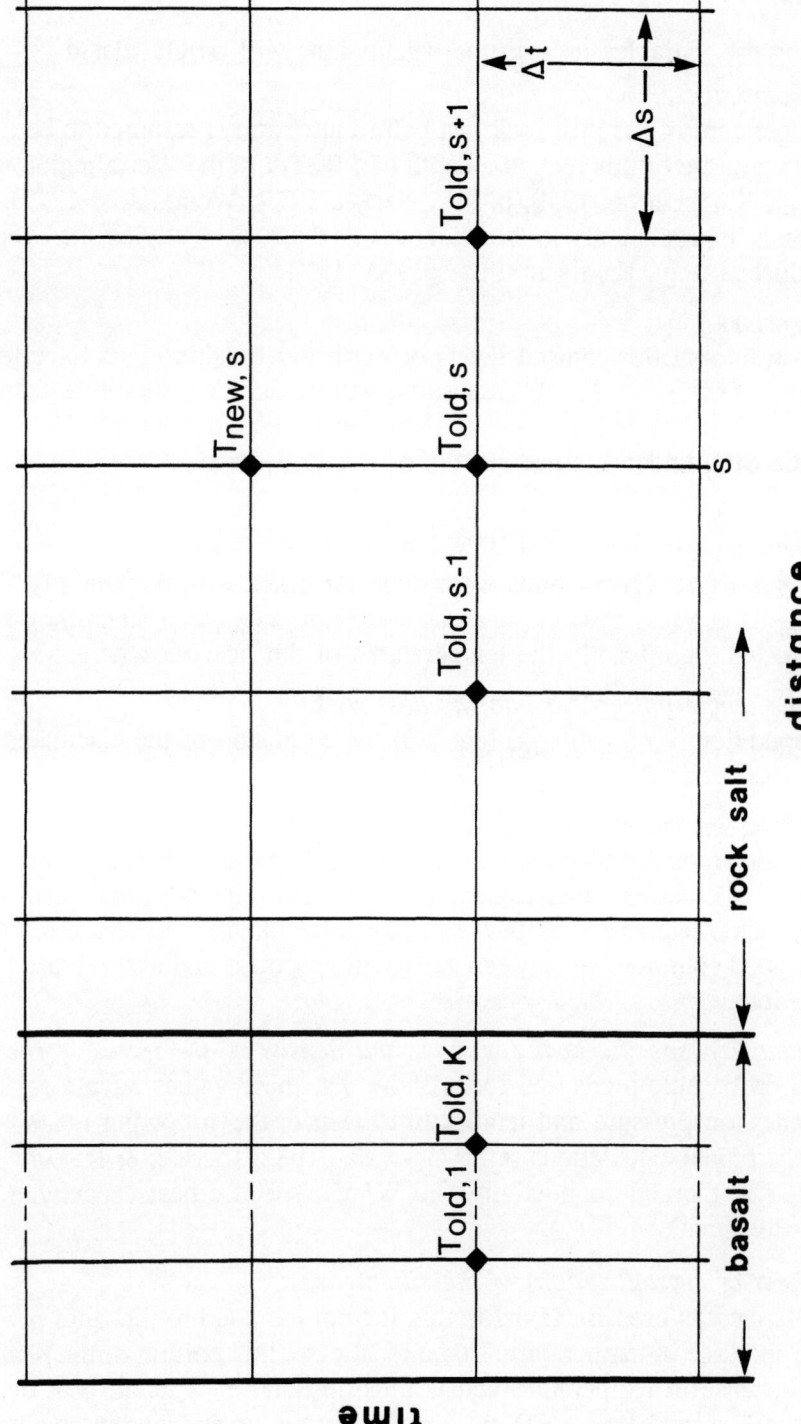

Fig. 8.1 Calculation of temperature $T_{new,s}$ based on the known T_{old} value (modified after SMITH 1965, 1970).

8.3 Variables

In the following the variables used in the calculations will be elucidated.

Host rock temperature
The initial temperature of the host rock just before intrusion is assumed to be 50 °C. The present temperature of this rock at a depth of 1 000 m in the Werra region is less than 50 °C (e.g., KAPPELMEYER 1959, p. 318; HAENEL 1979; WOHLENBERG 1979). This depth corresponds approximately to the thickness of the rock overlying the evaporite beds in the miocene (e.g., BRAITSCH 1962, p. 135, 1971, p. 177).

Magma temperature
During intrusion into shallow crustal levels or extrusion basaltic melts have temperatures of about 1 000 °C - 1 200 °C (measured values from the literature compiled by CARMICHAEL et al 1974). In the calculations presented here a value of 1150 °C was assumed for the SiO_2-undersaturated basalts of the study area.

Dike width
The basalt dikes exposed underground which were the objects of this study were about 0.05 - 1.8 m thick. Calculations were done for dikes with thickness of 0.3 m, 1 m, and 1.8 m. Dikes less than 0.3 m thick were not considered since they hardly could have affected significantly the temperatures of the host rock here.

Extension of the temperature field into the host rock
To meet the condition $T_{new,n} = T_{new,n-1}$ (eq. 8.3) the extension of the calculated temperature field was set at 20 m.

Distance and time intervals
In initial calculations the distance interval (Δs) was 0.01 m, and the time interval (Δt) 1 sec. To reduce calculation time, calculations were subsequently done with larger intervals. A Δt of 60 sec. and a Δs of 0.1 m (for a 0.3 m dike, 30 sec. and 0.05 m) proved to be a good compromise between sufficient exactness and expenditure for the model calculations.

Thermal conductivity, specific heat capacity, and density of the intrusive rock
Available data on basaltic melts and rocks show that these values have a relatively slight dependency on pressure and temperature. Hence, the following mean values were used in the model calculations (after GORANSON 1982; CERMAK & RYBACH 1982; WOHLENBERG 1982): thermal conductivity, 2.0 W/mK; specific heat capacity, 1.14 J/kgK; density, 2.9 g/cm³.

Latent heat (heat of crystallization) of the silicate melt
As a rock melts, heat is used for transforming it from the solid to the fluid phase, in addition to the increase in temperature. In the ideal case, this portion of the heat does not effect an increase in temperature and is thus designated as latent heat or (with constant pressure) as enthalpy (ΔH) of fusion. During crystallization the heat of fusion is released as heat of crystallization.

During intrusion, the heat of crystallization is surely distributed in the host rock over greater areas through the release of volatile phases and hence is only partially effective in the immediate vicinity of the contact (LOVERING 1936; JAEGER 1957). LOVERING (1936, 1955) suggested that the amount of latent heat in the magma be equated with exactly the amount of heat which is distributed over large areas of the host rock through the release of mobile phases from the melt. The latent heat could thus be disregarded since it would be compensated by the degassing of the fluid components of the magma. Yet, the extent of such processes cannot be measured directly. This is particularly true for an easily soluble host rock, like rock salt for example.

Various methods have been suggested and employed for calculating the latent heat (e.g., LOVERING 1936; LARSEN 1945; CARSLAW & JAEGER 1959; JAEGER 1964; JOBMANN 1985). In this study the following equation was used for approximation (after LARSEN 1945; cf. JAEGER 1957, p. 307 - 308):

$$c' = c + \frac{L}{T_i - T_s} \tag{8.5}$$

c' modified specific heat capacity of the intrusive rock in kJ/kgK
c mean specific heat capacity of the intrusive rock in kJ/kgK
L latent or crystallization heat in kJ/kg
T_i temperature of the silicate melt at the time of intrusion in °C
T_s solidus temperature of the magma in °C

In the theoretical case of a one-component system the latent heat is released during the crystallization of a silicate melt at a defined melting point. Natural basaltic melts are, in contrast, multicomponent systems. As a result, the latent heat is released over a wider temperature range. The increased specific heat capacity in a certain temperature interval is a measure of the enthalpy of fusion for indirect determinations with differential thermoanalysis (YODER 1976). In these model calculations the value for the specific heat capacity between intrusion and solidus temperature of the magma is modified with equation (8.5).

Voluminous data on the crystallizatian behavior of basaltic melts can be found in, for example, WYLLIE (1971), CARMICHAEL et al (1974), RINGWOOD (1975), or YODER (1979). According to RINGWOOD (1975, extrapolated) the solidus temperature of alkali-olivine-basaltic magmas is about 1 075 °C at atmospheric pressure. CARMICHAEL et al (1974, p. 7) compiled values from the literature on recent nephelinitic melts from the Congo and gives an intrusion temperature of 980 °C. The solidus temperature of the magmas in the model calculations is assumed to be 950 °C. Consequently, a specific heat capacity of 3.14 kJ/kgK can be calculated for the temperature range of 1150 °C - 950 °C with equation (8.5). The effect of the latent heat on the cooling path of the silicate melt in the model calculations is depicted schematically in fig. 8.2.

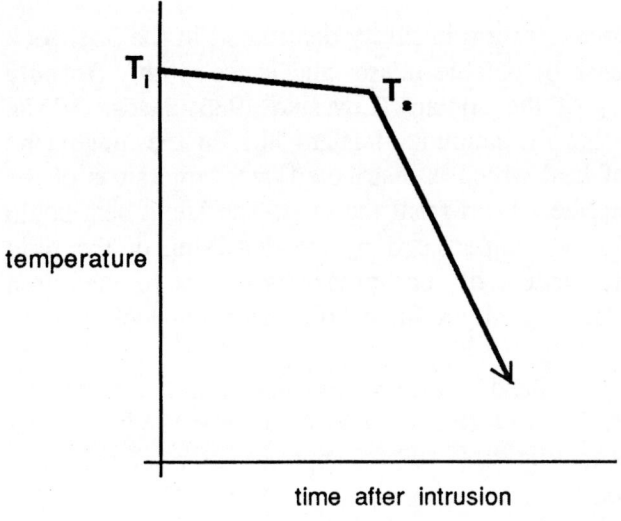

Fig. 8.2 Cooling path of a silicate melt allowing for the latent heat according to equation (8.5). T_i, intrusion temperature; T_s, solidus temperature. The initially slow cooling is attributed to the release of latent heat.

One difficulty is finding data on the absolute amount of the heat of fusion of a basaltic magma. The converted value of 620 kJ/kg melt experimentally determined by FUKUYAMA (1985) is valid for a 15 % partial melting of mantle rock. Based on the mineralogy and chemistry of the basalts from the Werra region, however, a partial melting of about 2 - 8 % of mantle rock is deduced (cf. WEDEPOHL 1983, 1985). Other data on the latent heat of basalts range from 300 to 400 kJ/kg (e.g., LOVERING 1936; JAEGER 1968; YODER 1975; JOBMANN 1985).

As the studies on their chemistry and mineralogy discussed in chapter 5 have shown, the basaltic magmas which intruded into the evaporites have variable amounts (vol%) of already crystallized minerals. While, for example, the basanites are very enriched in phenocrysts, the limburgites consist predominantly of glass. However, only a certain amount of the total heat of fusion is given off by a nearly completely crystallized magma. In the calculations presented here a value of 400 kJ/kg is used for the heat of crystallization.

Due to the unreliable data on the amount of latent heat and the simplified mathematical treatment the calculated temperatures for the interior of the basalt bodies are only reference points. Yet, JAEGER (1959, p. 45) points out that the calculation results for the temperature distribution in the host rock are realistic in spite of the idealization (eq. 8.5).

Thermal conductivity of halite
It was known already before the decisive experimental papers of BIRCH & CLARK (1940a, b) that the thermal conductivity of halite is strongly dependent upon temperature at < 400 °C. Due to its effect on the temperature distribution in the host rock this dependency should be taken into account in the model calculations. However, measured values for temperatures > 400 °C are not available.

SCHMIDT (1971) used a polynomial of the third degree for approximating thermal conductivity as a function of temperature. Nonetheless, examination of the coefficients below 400 °C yielded an unsatisfactory agreement with the measured values of BIRCH & CLARK (1940a, b), and from 400 °C to 800 °C with the curve path given by SCHMIDT (1971, fig. 2.03). To obtain valid function values for the range between 0 °C and 800 °C a polynomial of the fourth degree was chosen for this study. The values above 400 °C used were able to be extrapolated from the specific thermal resistance (reciprocal values of the thermal conductivity) by linear progression (table 8.1, fig. 8.3), as suggested by BIRCH & CLARK (1940a, b). In spite of the uncertainty involved with such an extrapolation it offers the best possibility of obtaining function values for temperatures above 400°C, apart from experimental determination. Calculation of the polynomial with the form

$$\lambda h = a_0 + a_1 T + a_2 T^2 + a_3 T^3 + a_4 T^4 + a_5 T^5$$

(λh = thermal conductivity of halite) yielded the following coefficients:

$a_0 = 6.113$
$a_1 = -2.366 \cdot 10^{-2}$
$a_2 = 5.525 \cdot 10^{-5}$
$a_3 = -6.483 \cdot 10^{-8}$
$a_4 = 2.904 \cdot 10^{-11}$

With the help of the coefficients the thermal conductivity at every T_{old} in the host rock can be calculated and thus can be used for calculating T_{new}. The path of the function values (fig. 8.4) in the range of 0 °C - 400 °C overlaps the values measured by BIRCH & CLARK (1940a, b) and the curve worked out by DELISLE (1980, p. 466).

Since BIRCH & CLARK (1940a, b) carried out their studies on individual halite crystals, the function in fig. 8.4 is valid, strictly speaking, only for halite. Nevertheless the measurements of CREUTZBURG (1965) on the *Älteres* and *Jüngeres* rock salt of the Stassfurt sequence for 35 °C show that the function is a good mean for natural rock salt.

Specific heat capacity and density of the host rock
The mean values for the specific heat capacity (0.88 kJ/kgK) and density (2.14 g/cm³) of rock salt were calculated from the data of CERMAK & RYBACH (1982) and CREUTZBURG (1965).

Table 8.1 Experimental (BIRCH & CLARK 1940a, b) and extrapolated (*) values used for determining the function of the dependency on temperature of the thermal conductivity of halite.

temperature [°C]	0	200	400	600	800
thermal conductivity of halite [W/mK]	6.113	3.119	2.085	1.574*	1.265*

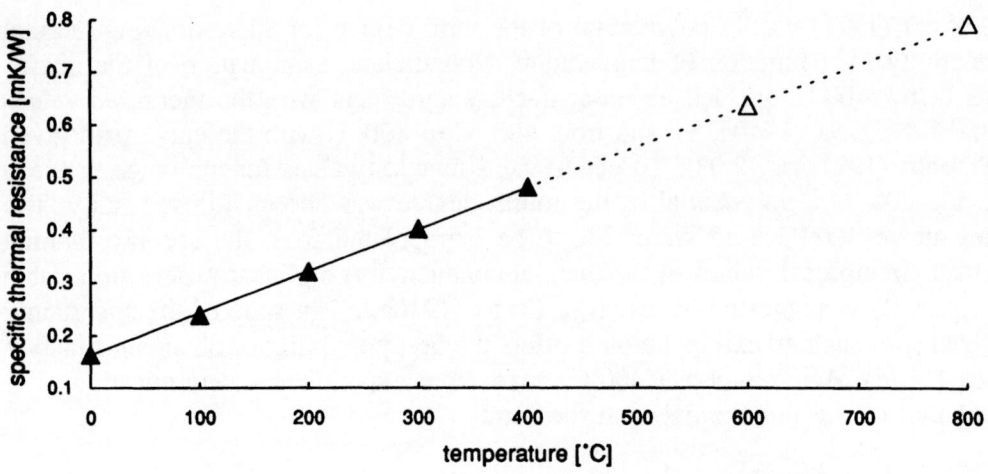

Fig. 8.3 Specific thermal resistance of halite (reciprocal value of the thermal conductivity). The values for temperatures above 400 °C (dashed line) are extrapolated.

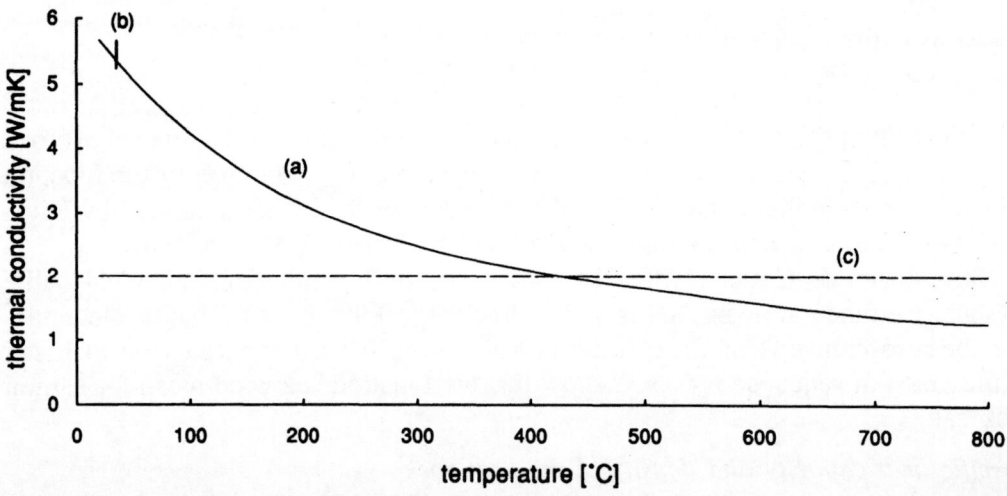

Fig. 8.4 The thermal conductivity of halite in dependency upon temperature. The path of the function (a) was calculated with a polynomial of the fourth degree. Results of measurements on natural rock salt (b) at 35 °C (CREUTZBURG 1965, p. 70) and the mean thermal conductivity of basalt (c) calculated from data in CERMAK & RYBACH (1982, p. 332) are also plotted.

The data used in the calculations are compiled once again for a better overview:

form of the intrusive body	dike
intrusive rock	basalt
host rock	rock salt
temperature of host rock	50 °C
temperature of magma	1150 °C
thickness of dikes	0.3 m, 1 m, 1.8 m
extension of the temperature field	20 m
distance intervals	0.05 m and 0.1 m
time intervals	30 sec. and 60 sec.
intrusive rock: thermal conductivity	2.0 W/mK
specific heat capacity	1.14 kJ/kgK
density	2.9 g/cm^3
latent heat	400 kJ/kg
host rock: thermal conductivity	f(T)
specific heat capacity	0.88 kJ/kgK
density	2.14 g/cm^3

f(T), function of temperature

8.4 Results

The temperature fields in the rock salt were calculated for three examples, i.e., for dikes with thickness of 0.3 m, 1 m, and 1.8 m for 0 to 365 days following intrusion of the basaltic melts. Due to partially unfavourable assumptions (chapter 8.2 and 8.3) the model might yield temperatures which are slightly too high. For the sake of clarity the results of the calculations are presented in diagramatic form. Important stages of the cooling and heating process in the basalt and at the contact of the three dikes are summarized in table 8.2. Since the values for the interior of the basalt dikes can only be used for reference, they are not depicted in the figures. The representation of the temperature field in the rock salt was able to be limited to a distance of 5 m from the contact because significant changes in temperature are only found near the intrusive dikes.

Immediately following intrusion of the silicate melts a temperature of about 600 °C prevailed at the basalt-rock salt contact (fig. 8.5). After about six hours the contact temperature reaches a maximum value of 790 °C (780 °C for the 0.3-m-thick dike). Up to this time the changes in temperature were nearly the same for all three calculated examples. When the dikes are thinner then 0.4 m, the heat contained in the melt is inadequate for heating up the rock salt at the contact to 790 °C (table 8.2).

As long as the temperature in the middle of the dike still remains above the assumed solidus temperature, latent heat is released. The reduction in temperature at the contact then takes a different temporal path dependent upon the thickness of the dike and thus the absolute mass of the melt (figs. 8.6 - 8.8). Cooling is more rapid directly at the contact of thinner dikes. Meanwhile the heat front wanders constantly

Table 8.2 Calculated (rounded off) temperatures [°C] in the interior of the basalt dikes and at the basalt-rock salt contact in dependency on the time following intrusion. The maximum temperature at the contact in each case is printed in bold type. Besides initial and final temperatures the points in time are shown at which the magma temperatures in the middle of the dike fell below the intrusion temperature of 1 150 °C in each case (see < 1 150). The total latent heat is used up when the temperature in the interior of the dike sinks below 950 °C (see < 950).

thickness of dike	0.3 m		1 m		1.8 m	
	middle of dike	contact	middle of dike	contact	middle of dike	contact
temperature before intrusion	-	50	-	50	-	50
after 1 sec	1 150	600	1 150	600	1 150	600
after 20 min	< 1 150	660	1 150	660	1 150	660
after 5.5*, 6 h	*1 070	*780	< 1 150	790	1 150	790
after 11 h	< 950	736	1 149	780	1 150	790
after 30 h	620	570	1 130	780	< 1 150	780
after 5 d	320	310	< 950	760	1 120	780
after 16 d	180	180	540	530	< 950	760
after 1 a	70	70	130	130	220	220

sec, second; min, minute; h, hour; d, day; a, year

further into the host rock. When the solidus temperature is reached over the entire width of the dikes, an accelerated cooling at the contact is observed (table 8.2; figs. 8.6a - 8.8a). Calculations for times of more than one year after intrusion were no longer done in view of the rapid cooling of the dikes under consideration.

Based on a simplified model of JAEGER (1957) WINKLER (1967) calculated the maximum temperatures in the unspecified host rock of a gabbro dike. The values calculated for the dike width of 1 m agree well with the results presented here (table 8.3). Although in this study a greater difference between intrusion and solidus temperature and thus a larger cooling time is assumed, the maximum temperatures are only negligibly higher. This can obviously be attributed to the thermal conductivity of rock salt which increases clearly below 400 °C (fig. 8.4). In addition, the dependency of the thermal conductivity on temperature leads to a shallower cooling curve with some distance from the contact. In contrast, a steeper cooling curve is observed for the host rock directly at the contact shortly after intrusion (figs. 8.6a - 8.8a). Figure 8.9 shows, for comparison, the results of a temperature-field calculation which was done with constant thermal conductivity for rock salt (λ = 2W/mK). The maximum contact temperature here amounts to just over 800°C. As expected, the cooling curve for the host rock is steeper (see figs. 8.7b and 8.9b).

In fig. 8.10 the results of a calculation which did not allow for the heat of fusion of the basaltic magma are presented. It is evident here that cooling begins nearly directly after a maximum contact temperature of less than 670 °C is reached.

Table 8.3 Comparison of calculated maximum temperatures at a gabbro dike (WINKLER 1967) and a basalt dike (this study, see also fig. 8.7a, b). The latent heat was taken into account in both models.

	WINKLER (1967)	this study
rock type	gabbro	basalt
width of dike	1 m	1 m
temperature of magma	1 200 °C	1 150 °C
liquidus temperature	1 050 °C	950 °C
maximum temperature at contact	775 °C	790 °C
maximum temperature 0.1 m away from contact	675 °C	680 °C
maximum temperature 0.5 m away from contact	460 °C	470 °C

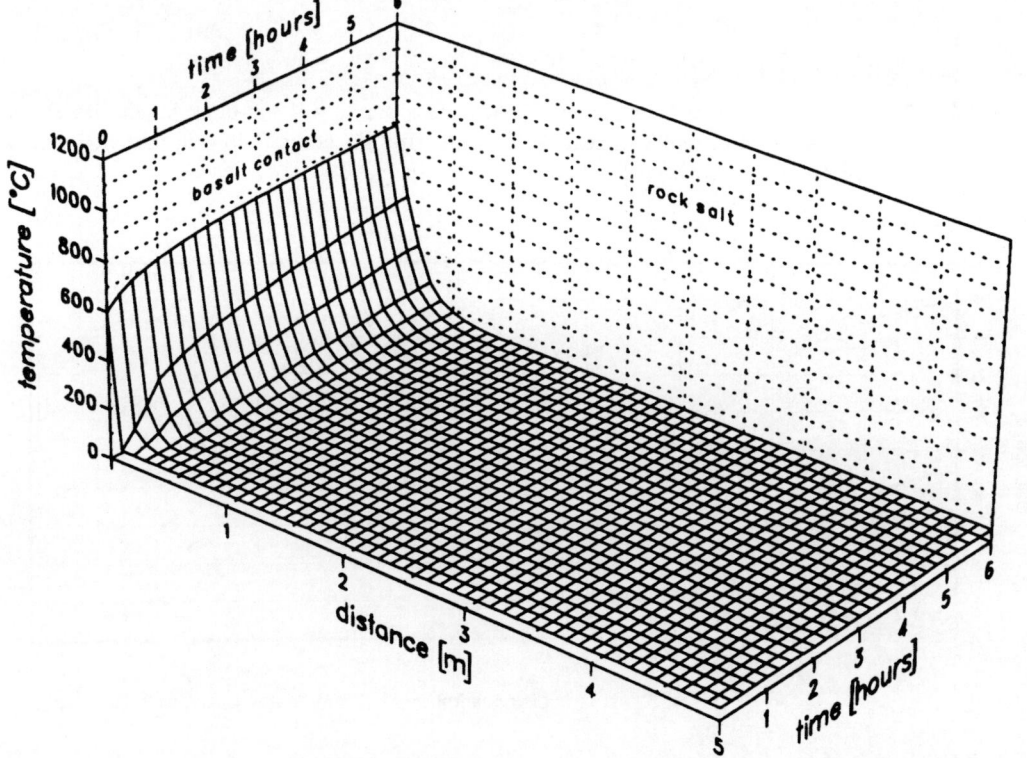

Fig. 8.5 Contact temperature and temperature field in rock salt at a 0.3- to 1.8-m-thick basalt dike in the first six hours following intrusion.

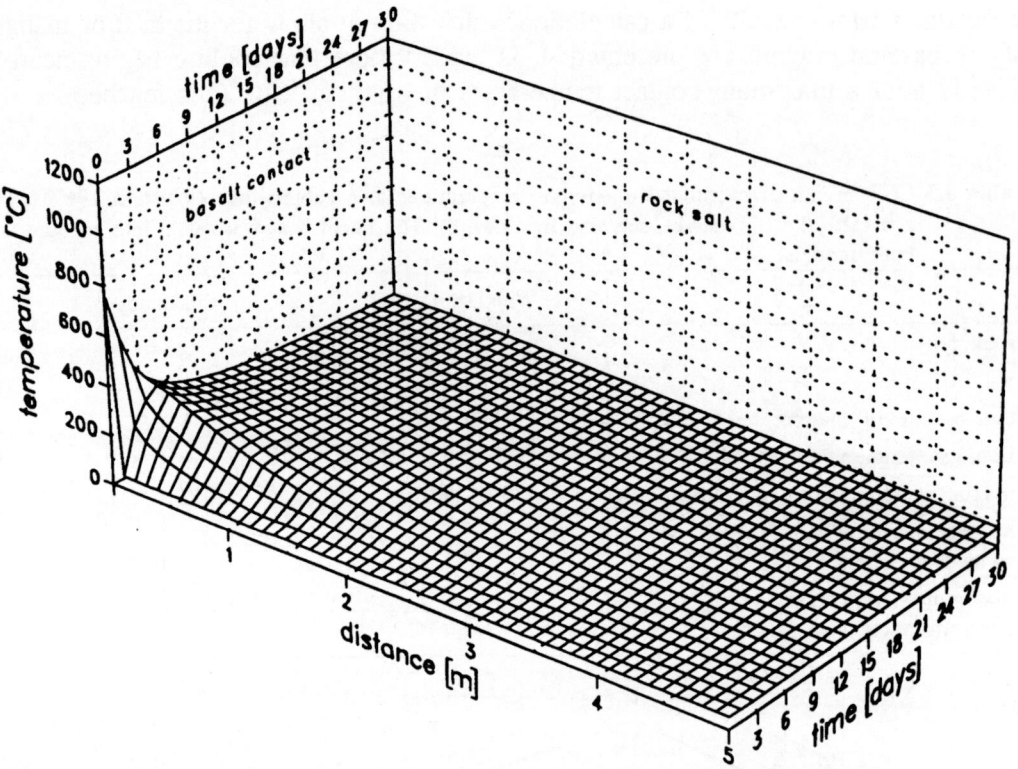

Fig. 8.6a Contact temperature and temperature field in rock salt at a 0.3-m-thick basalt dike 0 - 30 days following intrusion. The heating up in the first six hours following intrusion cannot be seen due to the scale and is hence depicted in fig. 8.5.

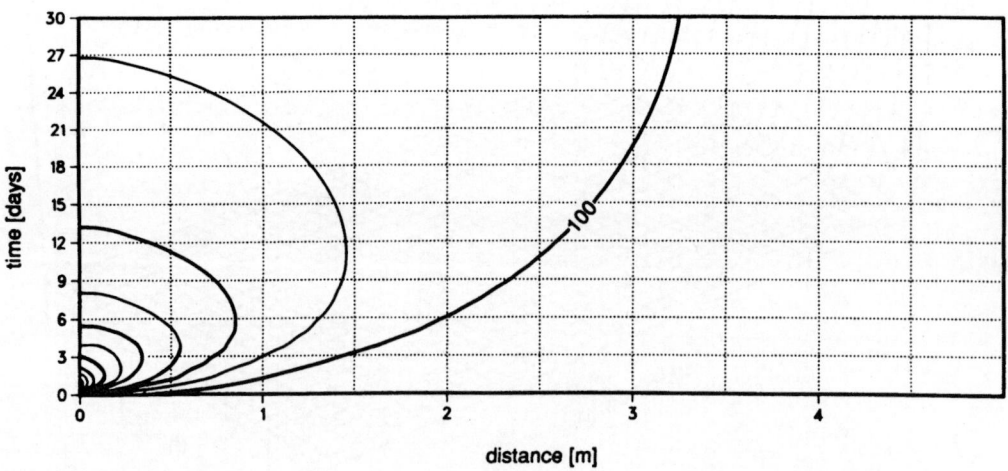

Fig. 8.6b Results from fig. 8.6a shown as two-dimensional isotherms.

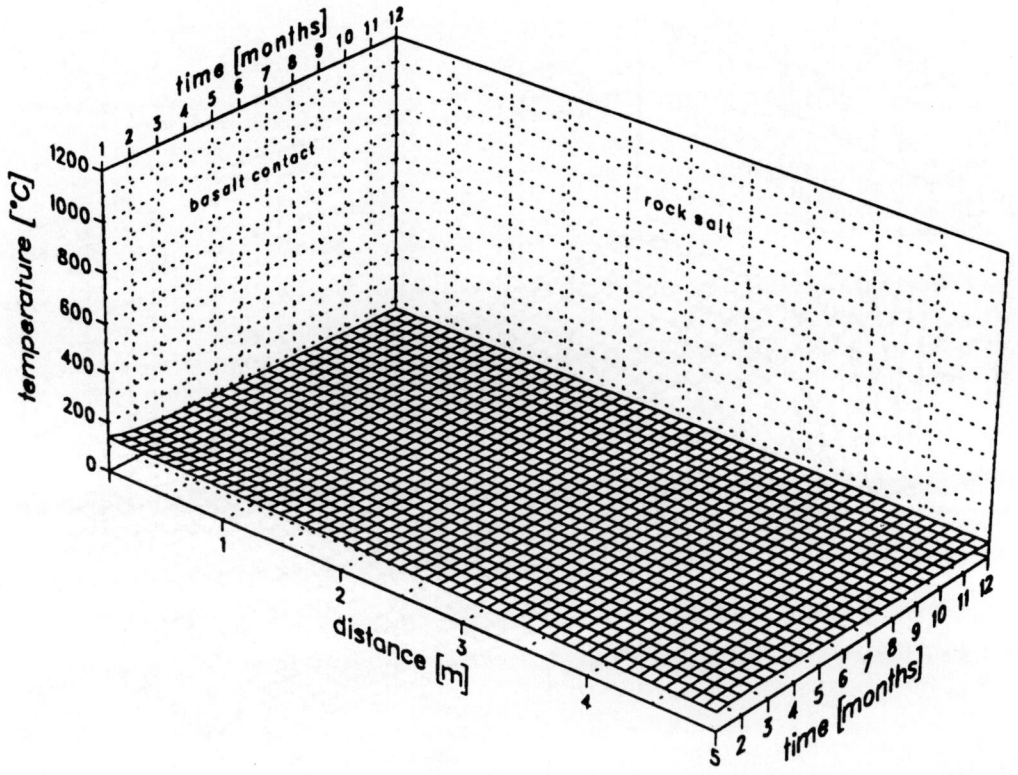

Fig. 8.6c Contact temperature and temperature field in rock salt at a 0.3-m-thick basalt dike 1 - 12 months following intrusion.

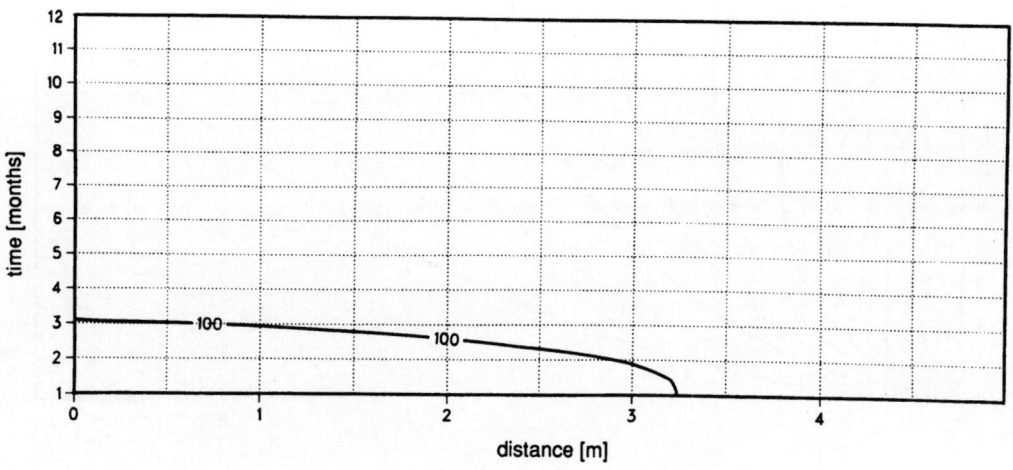

Fig. 8.6d Results from fig. 8.6c shown as two-dimensional isotherms.

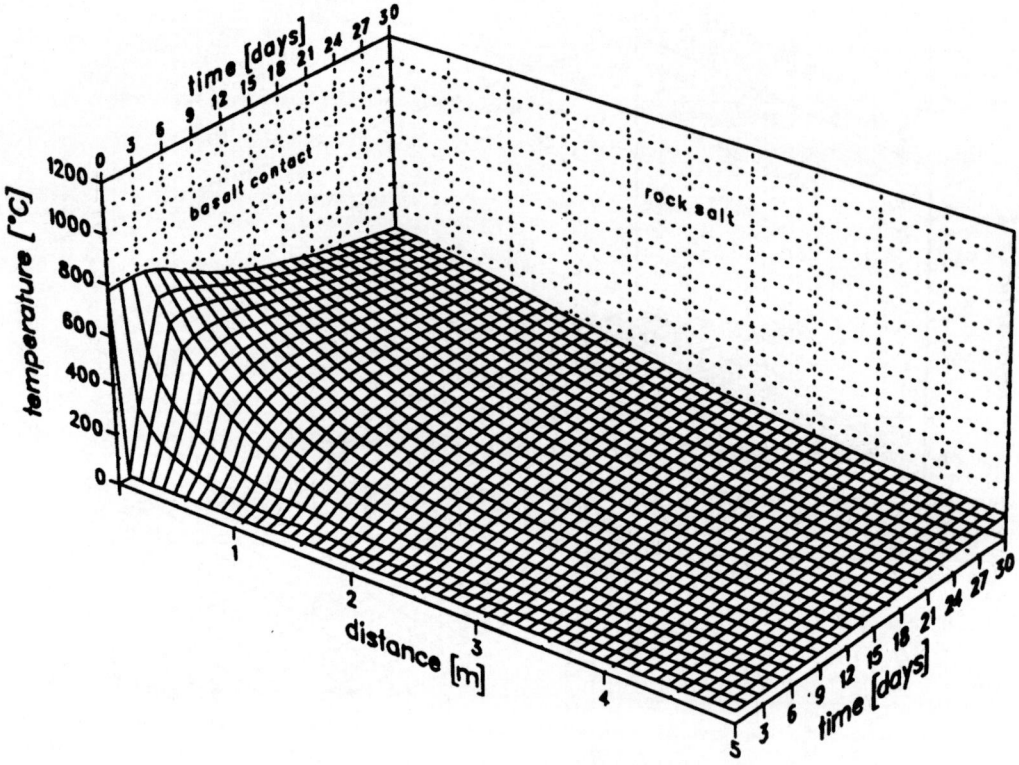

Fig. 8.7a Contact temperature and temperature field in rock salt at a 1-m-thick basalt dike 0 - 30 days following intrusion. The heating up in the first six hours following intrusion cannot be seen due to the scale and is hence depicted in fig. 8.5.

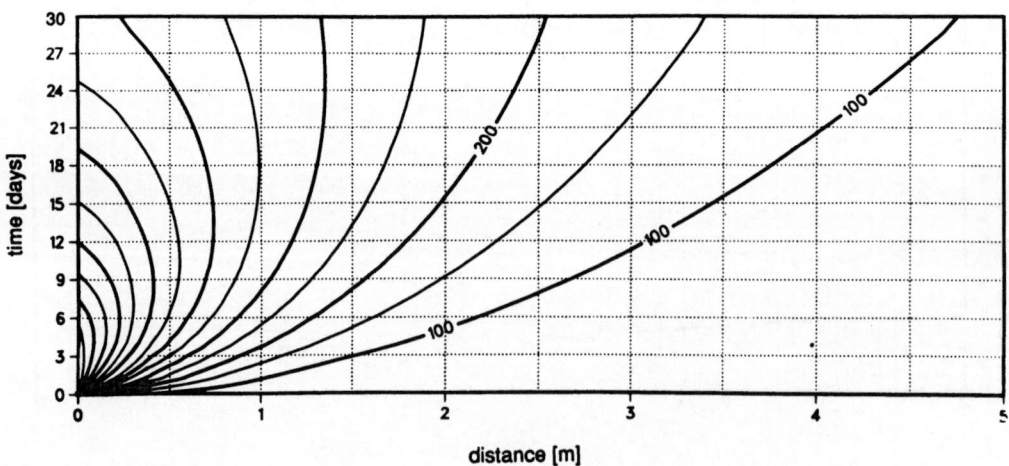

Fig. 8.7b Results of fig. 8.7a shown as two-dimensional isotherms.

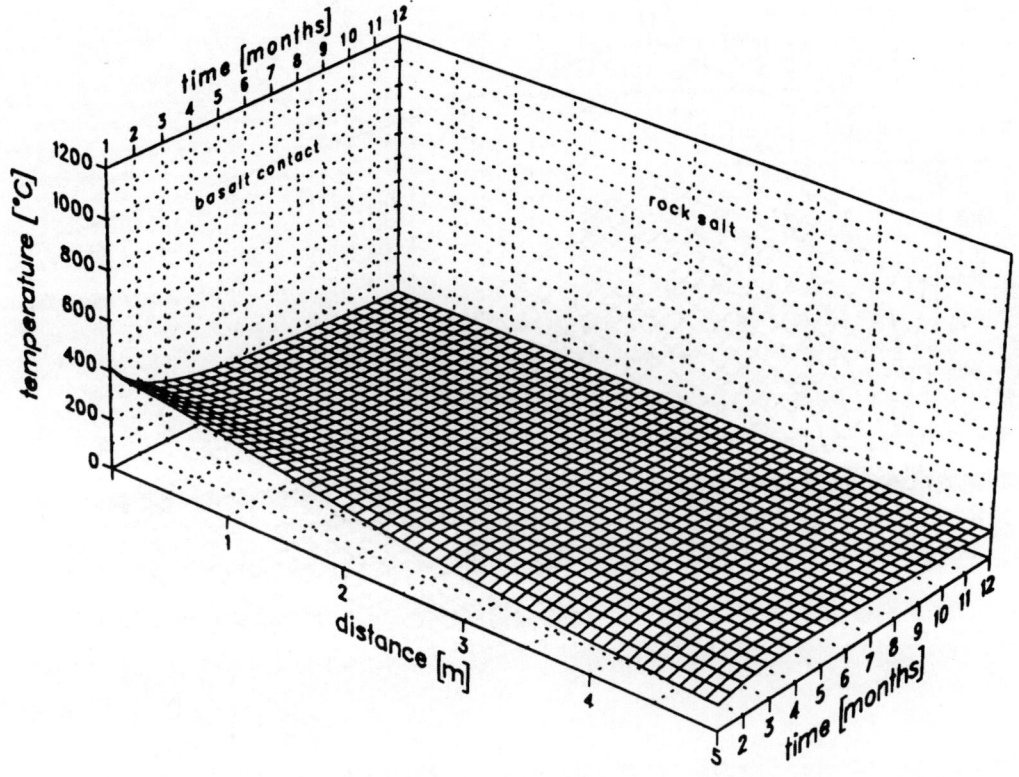

Fig. 8.7c Contact temperature and temperature field in rock salt at a 1-m-thick basalt dike 1 - 12 months following intrusion.

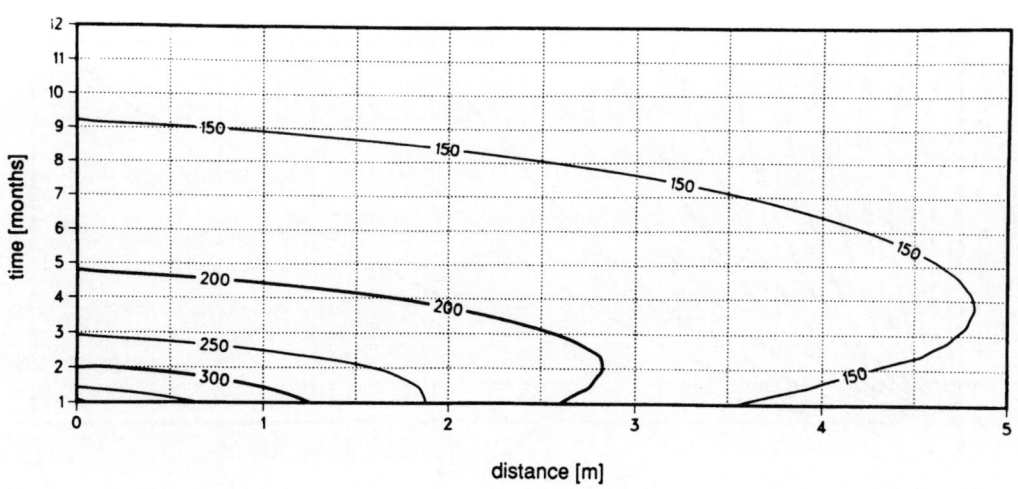

Fig. 8.7d Results of fig. 8.7c shown as two-dimensional isotherms.

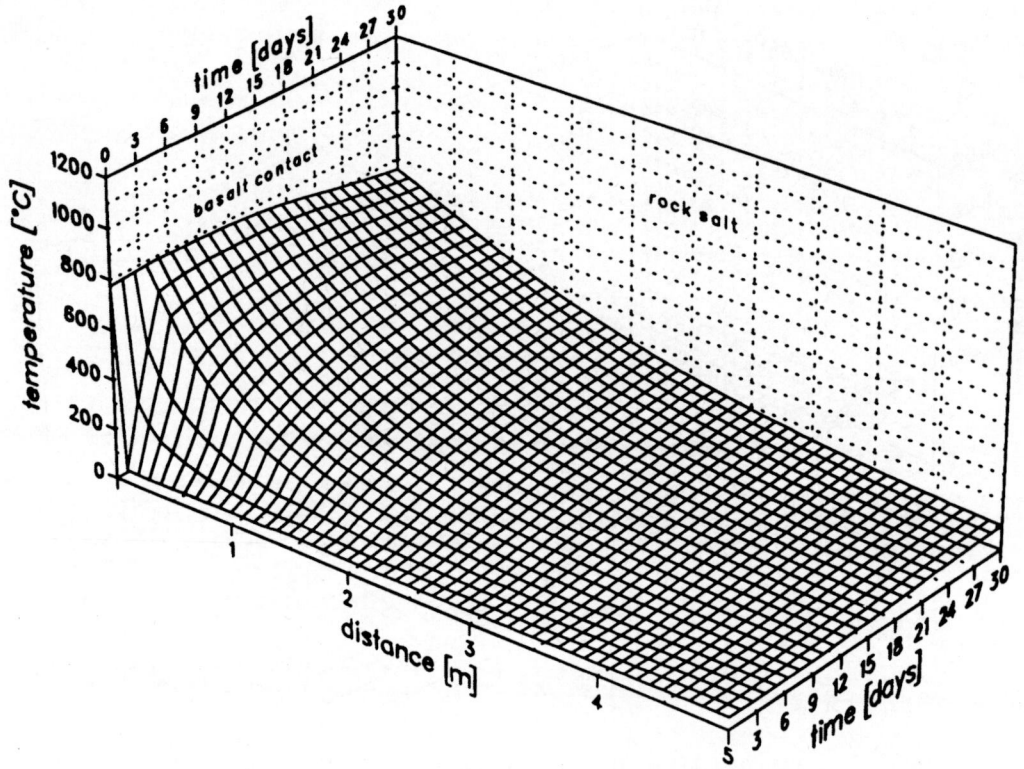

Fig. 8.8a Contact temperature and temperature field in rock salt at a 1.8-m-thick basalt dike 0 - 30 days following intrusion. The heating up in the first six hours following intrusion cannot be seen due to the scale and is hence depicted in fig. 8.5.

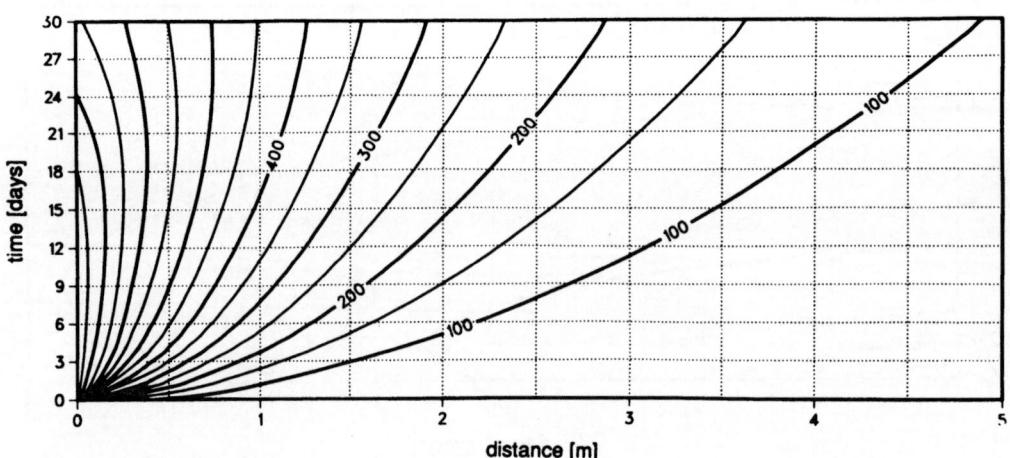

Fig. 8.8b Results from fig. 8.8a shown as two-dimensional isotherms.

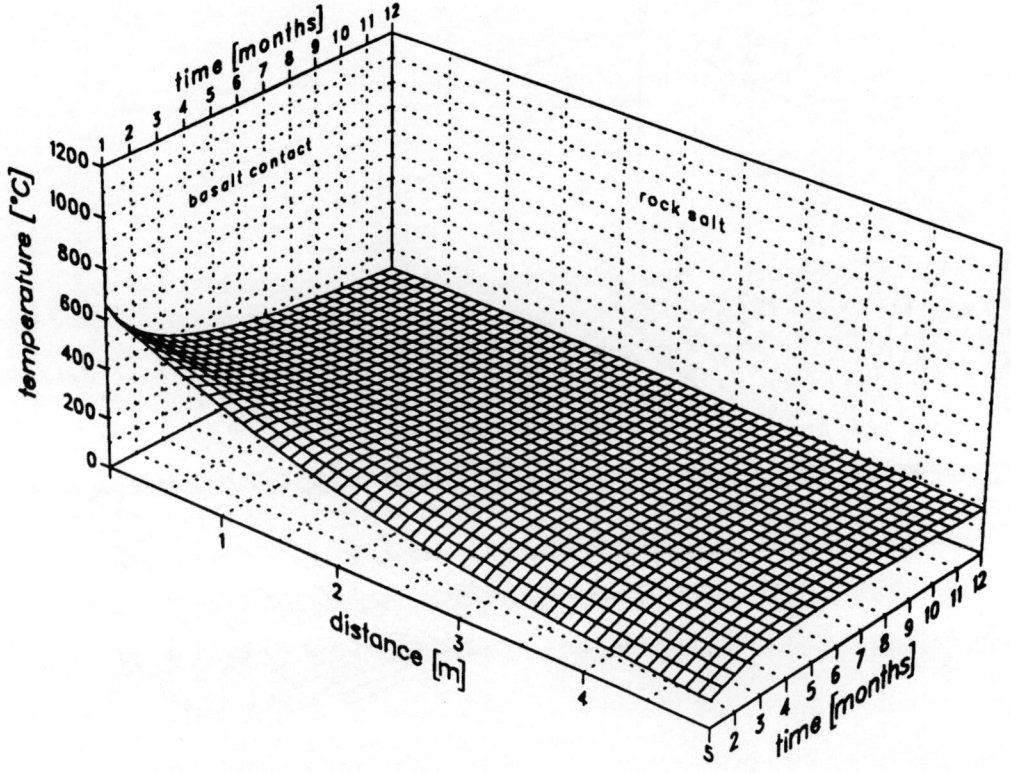

Fig. 8.8c Contact temperature and temperature field in rock salt at a 1.8-m-thick basalt dike 1 - 12 months following intrusion.

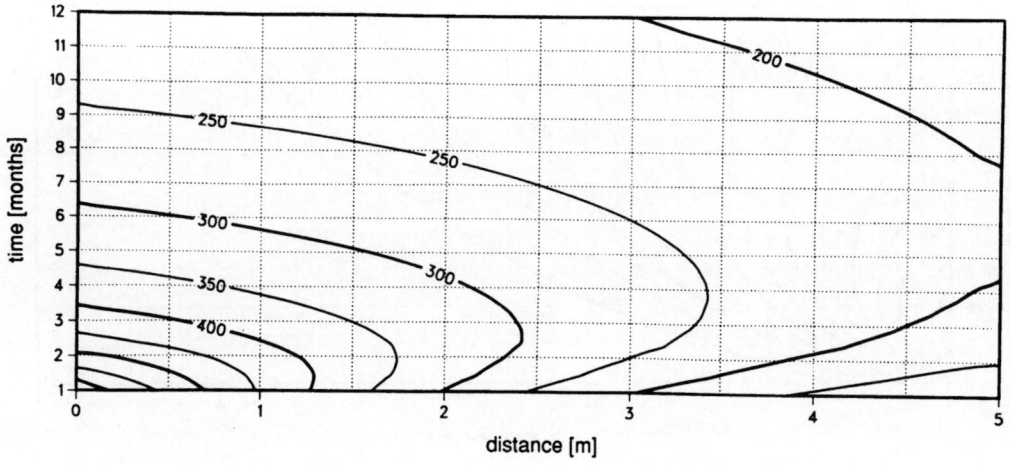

Fig. 8.8d Results of fig. 8.8c shown as two-dimensional isotherms.

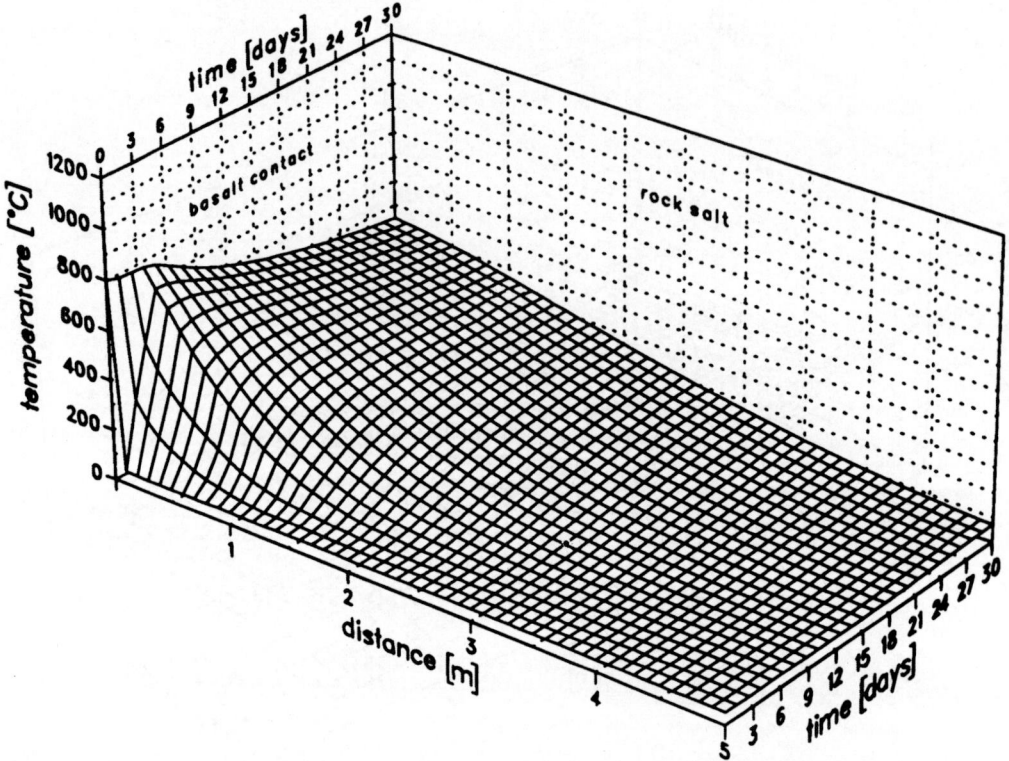

Fig. 8.9a Contact temperature and temperature field in rock salt at a 1-m-thick basalt dike 0 - 30 days following intrusion. Calculation with constant thermal conductivity for rock salt (2.0 W/mK). The heating up in the first 6 hours after intrusion cannot be seen due to the scale.

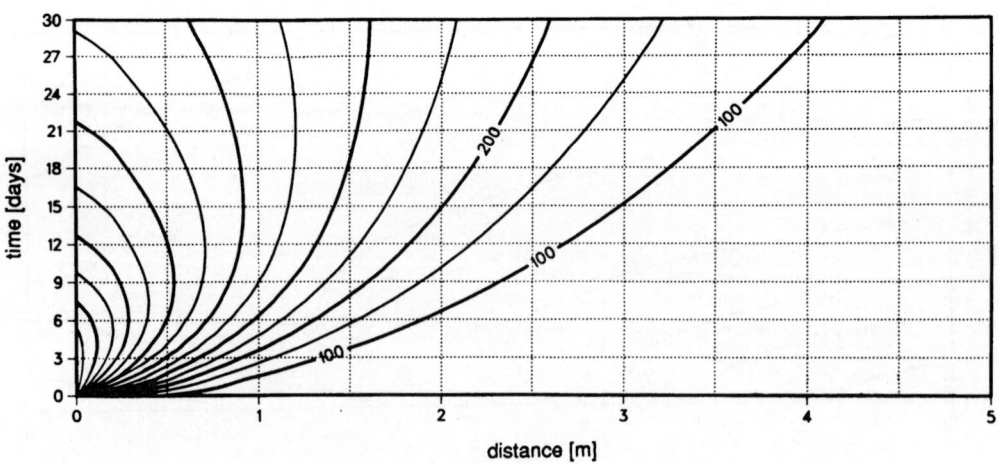

Fig. 8.9b Results of fig. 8.9a shown as two-dimensional isotherms.

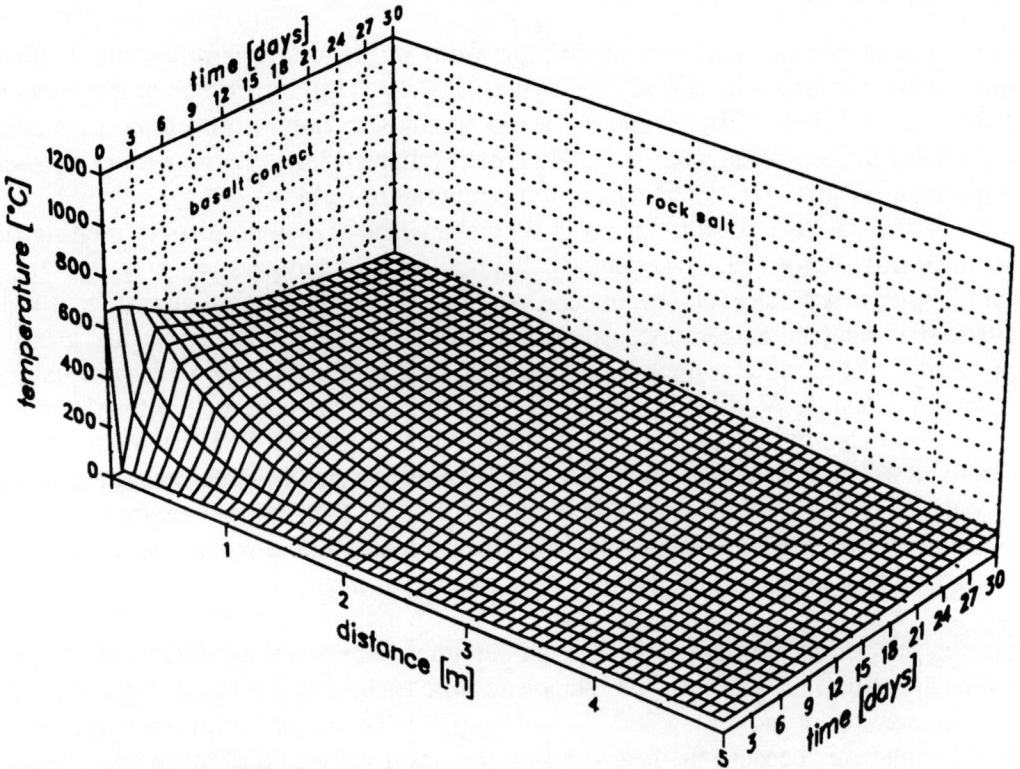

Fig. 8.10a Contact temperature and temperature field in rock salt at a 1-m-thick basalt dike 0 - 30 days following intrusion. Calculation does not allow for the heat of fusion of the magma. The heating up in the first 6 hours after intrusion cannot be seen due to the scale.

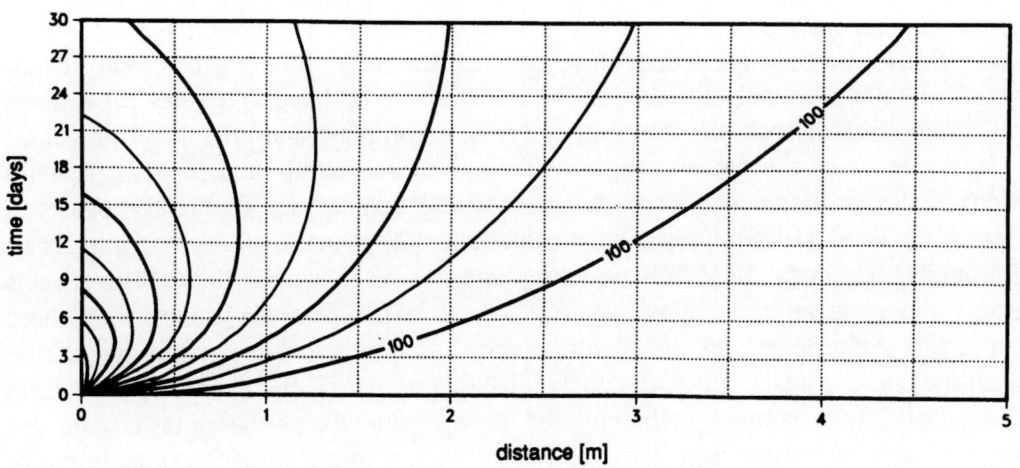

Fig. 8.10b Results of fig. 8.10a shown as two-dimensional isotherms.

8.5 Discussion

As the model calculations have shown, the terms *intrusion temperature* and *contact temperature* can by no means be used synonymously. The temperature at the contact to the host rock immediately after intrusion amounts to only a little more than half of the intrusion temperature of the silicate melt (chapter 8.4). The maximum contact temperature is just 70 % of the intrusion temperature (table 8.2).

Since the basalt-rock salt contact is very sharp and shows no distinct melting of the rock salt, KÜHN (1951) concludes that the magma temperature was less than 805 °C (due to this lack of pronounced contact zones). In contrast, KOCH & VOGEL (1980) state magma temperatures of clearly more than 800 °C. The 'typically glassy' appearance of the halite at the basalt contact is cited as proof of the melting of halite. However, KOCH & VOGEL (1980) do not distinguish between rock salt fragments which were transported by the silicate melt during intrusion and the halite at the margin wall of the basalt dike. The comparatively small halite xenoliths were surrounded on all sides by the silicate melt. Provided that the temperature remains the same for a rather long time in the interior of the basalt dike due to the latent heat, temperatures of clearly more the 800 °C could by all means have an effect on such rock salt fragments by melting the halite. Other conditions prevailed at the contact between the basalt and host rock. The temperatures here remained under 800 °C due to the dissipation of heat into the evaporitic host rock, and *the NaCl did not melt*.

Temperatures of more than 800 °C could possibly be attained between two closely neighboring dikes because the heat fronts in the host rock would advance toward each other (e.g., JAEGER 1968, fig. 2b). Furthermore, similarly high temperatures could occur in the acute angle between a vertical dike and an offshoot. With respect to the model calculations there is a qualitative observation that such phenomena are attached to several conditions. Parallel dikes must neighbor very closely and have a certain minimum separation. Otherwise, the dikes would already cool before the minimum temperature of the host rocks exceeds 800 °C. Offshoots from the main dikes with thicknesses great enough to produce the long cooling times likely occur extremely seldom as well. In addition the intrusion would have to be 'dry'. It follows that the melting of rock salt may have been a subordinate and only local process.

In the level of the potash seams widenings of the basalt bodies with partially extensive impoverishments on both sides of the contact are frequently observed (chapter 2). DIETZ (1928) described the more than 100-m-thick, sill-shaped expansion of a basalt body in the K1H horizon at the Sachsen-Weimar mine (now Marx-Engels mine). Frequent attempts have been made at explaining the genesis of these horizontal expansions without quantitatively interpreting the mineralogy at the basalt-evaporite contact. KOCH & VOGEL (1980) postulated, for example, the ascent of a premagmatic solution, whereby the K-Mg minerals of the potash seam are supposed to have existed as a 'pasty salt mass' before intrusion of the basaltic melts. However, these ideas cannot be correct for the following reasons:

- ascent of the magmas directly after the formation of paths by hydrofracturing (SHAW 1980),
- high ascent velocity of the basaltic magmas from the earth's mantle or an intermediate magma chamber (chapter 5 and 9),
- frequently higher mobility of the silicate melts due to enrichment of fluid phases (see SPERA 1980).

As a reason for the widening of the basalt, for example, in the kieseritic Hartsalz of K1H KÜHN (1951, p. 104) assumes that '...the evaporites were melted or reacted on one another...'. Kieserite releases crystal water at about 360 °C (D'ANS 1933, p. 118) or 335 °C (JOCKWER 1981). KÜHN also presumes that the salt minerals of the Hartsalz underwent a thermal metamorphism in solid form in the sense of LEONHARDT & BERDENSKI (1949/1950). On the other hand, BRAITSCH (1962, p. 88, 1971, p. 113/114) points out with regard to the works of these authors that solutions always form during thermal metamorphism.

The quantitative interpretation of the mineralogy of the salt rocks at the basalt contact can provide evidence of a thermal metamorphism with the help of the solution equilibria of marine evaporites. According to the studies of KNIPPING (1984) and KNIPPING & HERRMANN (1985) it has to be presumed in this case that the melt with the participation of higher-temperature aqueous solutions actively made way at a basalt-carnallitite contact. These processes then are a *solution metamorphism at increased temperatures* (not a thermal metamorphism). Horizontal basalt bodies like those exposed in the potash horizons have not yet been observed in the rock salt beds. The reason for this could be that the aqueous phases were completely saturated with NaCl during ascent through the Unteres and Mittleres Werra rock salt (cf. HARTWIG 1954, p. 8; BAAR 1958, p. 145; chapter 5.3). In the horizons of the potash seams these NaCl-saturated solutions then react with the K-Mg rocks.

BRAITSCH et al (1964) provided the first plausible evidence of the temperatures which were effective at the basalt-evaporite contact in their studies on datolite at the contact to an ankaratrite (melanocratic olivine nephelinite) of the Tertiary salt deposits on the Upper Rhine (Buggingen potash mine). Based on the occurrence of datolite in the form of small nodules the authors conclude that this boron mineral must have grown quickly as a result of the rapid cooling at the contact. They calculated contact temperatures, i.e., a formation temperature for datolite, of around 300 °C from the transformation of the original sylvinite into a halite rock (impoverishment) which obviously occurred without any substantial change in volume.

LOEHR (1979) described two, 5- and 1.4-m-wide 'lamprophyr' dikes of alkali-olivine-basaltic composition in the Permian evaporites of the Salado Formation near Carlsbad, New Mexico. Unfortunately, LOEHR (1979) has frequently left out clear data on the kind or type of the salt rocks. The age of these basalt dikes is reported to be Tertiary. The fluid inclusions contained in the neighboring salt rocks are supposed to have been affected only slightly by the basaltic melts in an undescribed way (LOEHR 1979). ROEDDER & TAGGERT (1978, cited in LOEHR 1979) determined homogenization temperatures of 71 °C - 150 °C for the fluid inclusions with bubbles in the rock salt at the basalt contact. LOEHR (1979) regards these temperatures as being too low since she assumes that the salt was melted up to 15.2 cm (!) from the basalt contact. However, this assumption is speculative. BROOKINS (1986) presumes an intrusion temperature of about 850 °C for the same dike and writes '...Only in the immediate contact zone of a few centimeters were there any pronounced effects, i.e. partial melting of halite with new salts formed. A new generation of polyhalite formed in the contact zone...'. These statements are also unfortunately based on speculation. The mineralogicial composition was not quantitatively interpreted with the aid of reaction paths. It is still unclear what BROOKINS (1986) meant with '...partial melting of halite...' and how the NaCl-free mineral polyhalite is formed by the melting of halite.

One occurrence of basalt in K1Th in which relictic carnallite was observed even up to 2 cm away from the basalt contact was described by KNIPPING (1984) and KNIPPING & HERRMANN (1985). KERN & FRANKE (1980, fig. 5) state that the incongruent decomposition (melting) of carnallite begins at 8 MPa at about 150 °C. In certain zones surrounding the basalt body an incongruent alteration of carnallite through hot solutions as well as a renewed crystallization of carnallite with falling temperature were shown (KNIPPING & HERRMANN 1985). The aqueous solutions obviously penetrated into the salt rock very rapidly and under high pressure so that there was not sufficient time for the carnallite to be completely decomposed. The possible effects of a thermal metamorphism were superimposed by the processes of the solution metamorphism. The presence of the mineral carnallite at the basalt contact indicates that following intrusion of the silicate melts a continuously stronger heating of the salt rock did not occur. This could be further evidence for the rapid cooling of the silicate melts through the transport and distribution of heat by mobile phases.

The recent exposure of further basalt outcops in the Trümmer carnallitite (K1Th) of the Hattorf mine demonstrate that these observations are not isolated cases. Here as well, carnallite occurs again in the immediate vicinity of the basalt contact (v. BORSTEL 1989). KOCH (1978), KOCH & VOGEL (1980, fig. 3), and BROOKINS (1984) also reported the occurrence of carnallite at the contact to the intrusive rock.

BRAITSCH (1962, p. 135, 1971, p. 177) points out that during the time of active volcanism in the Tertiary host rock temperatures distinctly higher than 50 °C could possibly have prevailed due to an increased heat flow in the basement rock in the Werra region. The highest subsurface temperatures in the Federal Republic of Germany today were measured in the Urach volcanic region and in the Pfalz (Landau). At 1 000 m depth they amount to about 80 °C - 100 °C (HAENEL 1979), but are caused by water circulating deep in the basement rock. Structural joints possibly served as paths for the circulating water. In spite of the Quaternary volcanism in the Eifel (FUHRMANN & LIPPOLT 1982) the geothermal gradient is presently normal there (DELISLE, personal communication, 1987). Hence, it is rather unlikely that the Miocene volcanism in the area was responsible for a widespread heating up of the Werra region.

In summary, the model calculations for the intrusion of 'dry' basaltic melts gave an impression of the temporal and spatial temperature distribution in the hosting rock salt beds. Since the calculations showed that the contact temperatures remained under 800 °C, the halite there did not melt. NaCl can only have melted locally (xenoliths, branching dikes). Moreover, many observations evidence that the invasion of mobile phases into the evaporites related to the intrusion of silicate melts must have been the determining factor for the spatial and temporal temperature distribution at the basalt-rock salt contact and in the rock salt. In addition to the rapid dissipation of heat over larger parts of the evaporites contingent on this, the temperature of the host rock in the immediate contact zone was able to be increased through pure thermal conductivity (if need be; figs. 8.6 - 8.8). More than 10 m away from the basalt contact the temperature increased only by 0.1 °C - 10 °C. In view of the fact that the basalt magmas intruded at different times the temperatures produced by the volcanism thus had only a slight effect on the evaporites.

9 Conclusions

In the course of Miocene volcanism volatile phases like H_2O, CO_2, and native sulfur migrated into the evaporites of the Werra region. The salt rocks and particularly the K-Mg rocks of the potash horizons incorporated these volatile components. Thus, they are partially accessible for direct study. The work presented here demonstrates that investigations of the interactions between basalts and evaporites as well as of the composition of the mobile phases of fossil continental basalts can help in broadening our knowledge of the genesis not only of the basalts but also of the evaporites. In addition, the study of intrusive mechanisms related to the deposition of heat-producing, anthropogenic wastes is of increasing significance (natural analogues; e.g., BROOKINS et al 1983).

KOCH (1978) as well as KOCH & VOGEL (1980) postulated that aqueous solutions ascended from deeper sedimentary strata before the intrusive events. As the 'dry' basaltic melts then penetrated into the level of the potash horizons, the salt rocks are supposed to have had a 'pasty' consistency. In doing so, the magmas, which the authors in an outdated conception derive from an olivine-basaltic parent magma, are supposed to have absorbed water in the zone of the potash horizons. These ideas were not confirmed by the studies presented here. In the following, qualitative conceptions of the intrusive mechanism of the basaltic melts are developed, which are oriented toward the results of this study.

During the Miocene the Hessian Depression was the site of intensive volcanism caused by global plate tectonic processes. The intracontinental tectonic stress field which led to the ascent of basaltic melts in the Vogelsberg and Rhön regions was effective up into the northern foreland. In the region of the Werra salt mining district fluid-rich melts, which formed at depths of about 70 - 90 km, reached up to the earth's surface at several places within hours or days. An example of this is the peridotite-xenolith-bearing olivine nephelinite of Soisberg (outcrop 42-sf). However, part of the magmas also stagnated at greater depths, forming slowly cooling magmatic bodies. Early crystallized minerals like olivine and pyroxene were separated from the residual melt by gravitational differentiation. This process led to a change in the initial composition and to a relative enrichment of the volatile components like H_2O and CO_2 in the resulting melt. So, for example, the hornblendes observed in the basanites and phonolitic-tephrites could have formed by partial recrystallization of existing pyroxenes. The stagnation time of the basaltic magmas at great depths could amount to > 300 000 years, as the case of the formation of a phonolitic tephrite (dike <O>) from an alkali-olivine-basaltic parent magma demonstrates.

The enrichment of H_2O in the stagnating basaltic melts decreases the viscosity of the melts. A 'free' gas phase may also have formed when the solubility of H_2O and/ or CO_2 was exceeded. The increase in pressure led to the formation of additional fissures along zones of weakness (i.e., *hydrofracturing*). The further exsolution of volatile phases in a shallower crustal level accelerated ascent further and led to the explosive eruption of basaltic magmas in some places at the earth's surface, consequently producing tuffs (i.e., basanite at Trumbachsköpfchen, outcrop 44-sf).

The processes just described also led to the formation of joints and fissures in the evaporite layers. In doing so, the rock salt strata were penetrated completely by the basaltic melts producing a sharp contact between the dikes and the host rocks (at the present outcrop level). In contrast the K-Mg rocks of the potash horizons, which are especially sensitive to increased temperatures and unsaturated aqueous solutions, reacted immediately with the saturated NaCl solutions transported by the basaltic melts. The high pressure present during intrusion of the magmas is documented here by (i) the occurrence of relict carnallitite at the basalt contact (KNIPPING & HERRMANN 1985) and (ii) the occurrence of native sulfur up to 40 m away from the basalt contact (chapter 6). The majority of the magmas did not reach the surface due to the release of pressure and relatively small quantities of melt.

The explosive intrusion of the melts as well as the alteration of the evaporites could also be attributed to the participation of 'external' water. So-called phreato-magmatic explosions can result when ascending magma comes in contact with groundwater (e.g., HOFMANN et al 1987). The thin basalt dikes at the Hattorf mine are in part very vesicular, which rather would be more indicative of an explosive exsolution of volatile phases than contact with water in underlying sedimentary strata. Yet, in view of the in part extensive alteration zones in the salt rocks it cannot be excluded at present that aqueous solutions subsequently ascended after intrusion of the melts as well. Further investigations may provide evidence of these processes.

The conceptions of the intrusive mechanism of the basaltic melts presented above will surely be modified by future studies; yet they do form the basis for further studies of the geologically interesting coexistence of evaporites and basalts. The very similar compositions of the basalts generally show that they also formed under very similar conditions. In summary, it can be stated based on these results that the alterations caused by the basaltic melts and the material exchange between the evaporites and basalts depend primarily on the following factors:
- the composition of the basaltic melts,
- the degree of fractionation of the melts and thus the relative enrichment of the volatile phases,
- the amount of melt,
- the pressure at the time of intrusion,
- the chemistry and mineralogy of the host rock,
- the depth at which the basaltic melts come to a standstill, and
- the extent of transport of solutions from the underlying strata.

In comparison to the increase in temperature in the salt rock due to pure heat conduction, the mobile components which penetrated into the evaporites during volcanism dominate in the alteration and material transport processes. Yet, in spite of the considerable stress experienced by the evaporites in the Werra region due to the processes described, the deposit is still one of the most productive in the world today.

10 Suggestions for further studies

This work may serve as the basis for further geological and mineralogical studies concerning questions on the mechanism of intrusion of basaltic melts and on quantifying mineral reactions and material transports in evaporites. Furthermore, the study of several of the described aspects is of interest for questions concerning the final storage of heat-producing, toxic wastes.

Basalts

More age determinations on the freshest surface basalts possible would aid in reconstructing the intrusive sequence of the melts in the Werra region. On this basis, it would be possible to improve the previous estimation of the temperatures experienced by the entire evaporite sequence during Miocene volcanism. In this context an attempt at dating the subsurface basalts with other methods of age determination would also be very important.

Microscopy of hornblendes in the phonolitic tephrite and basanites as well as quantitative analysis with the microprobe of samples from profiles through the width of the dikes may provide further information on the genesis of amphiboles. Considerable attention should be paid to petrological aspects to be able to say something about the depth of formation.

Further information on the temperatures effective in the contact zone and on the type and extent of mobilization of trace elements from the basalt can be expected from the direct study of neogenic minerals at the basalt-evaporite contact. The clay mineral horizons in the evaporites ('Tonlöser') should be included in such a study.

Evaporites

The systematic study of the gases trapped in the evaporites at the basalt contact and the comparison of these gases with those contained in basalt-free parts of the deposits offer the opportunity for expanding our knowledge on the mobile phases of fossil basaltic melts. A thorough geochemical and petrographical study of the reservoir rocks of the fluid phases may help in prognosing gas occurrences in areas of the deposit not yet exploited.

A direct study of the sulfur-bearing alteration zones involves the chemical and petrological investigation of the evaporites including the separate determination of sulfur isotope distribution in the minor portions of sulfide. This would also have to entail isotope-geochemical studies of the opaque minerals in the basalts. The development of isothermal and polythermal reaction models would provide information for understanding the formation of the sulfate enrichments (anhydrite, kieserite, polyhalite, kainite, langbeinite) near the basalts or at the basalt contact.

$\delta^{18}O$ and δD determinations of the salt minerals containing crystal water may provide evidence of the extent of the participation of water from deeper sedimentary strata in the alteration of the evaporites. In this context the geochemical and petrographical study of extensive impoverishments correlatable with the basalt volcanism would also be important. Further evidence of the effects of the metamorphic events on the salt deposits could be obtained by the chemical and isotope-geochemical study of the fluids contained in the salt rock strata. The microscopic fluid inclusions, the neogenic minerals in the joints and fissures, and abolute age determinations should also be included in such studies.

A reliable quantification of mineral reactions and material transports in the geological past is only possible based on thorough mineralogical and chemical studies of the solid, liquid, and gaseous components of an evaporite sequence. The sulfate type potash salts of the Werra-Fulda mining district (the sulfate type of marine evaporites collectively make up only 0.5 - 5 % of the marine evaporites worldwide; HERRMANN 1987, 1989) are particularly suitable for these calculations. Consequently, the aim of the further studies suggested here will be the following: On one hand, our knowledge on the mobile components of fossil continental basalts will be broadened. On the other hand, under the subject of *Quantifying the Genesis of Evaporites* and based on thorough regional studies an overall spatial and temporal picture of the alteration of entire evaporite bodies, which has not yet been attempted in this form, will be developed.

11 References

ACKERMANN, G., SCHRADER, R., HOFFMANN, K. (1964): Untersuchungen an gashaltigen Mineralsalzen, II. Teil: Methodik und Ergebnisse der gasanalytischen Untersuchungen. - Bergakademie, *11*: 676-679.

D'ANS, J. (1933): Die Lösungsgleichgewichte der Systeme der Salze ozeanischer Salzablagerungen. - 254 S., Berlin (Verlagsges. für Ackerbau).

D'ANS, J. (1967): Das CO_2 in Kalisalzlagern, sein Zustand und die Bedingungen seines Entstehens. - Kali u. Steinsalz, *4*, H. 12: 396-401.

D'ANS, J.; LAX, E.; SYNOWIETZ, C. (1967): Taschenbuch für Chemiker und Physiker, 3. Aufl., Berlin-Heidelberg-New York (Springer).

AOKI, K. (1963): The kaersutites and oxykaersutites from alkalic rocks of Japan and surrounding areas. - J. Petrol., *4*: 198-210.

BAAR, A. (1958): Über gleichartige Gebirgsverformungen durch bergmännischen Abbau von Kaliflözen bzw. durch chemische Umbildung von Kaliflözen in geologischer Vergangenheit. - Freiberger Forschungshefte, *A123*: 137-159.

BAERTSCHI, P. (1952): Die Fraktionierung der Kohlenstoffisotopen bei der Absorption von Kohlendioxid. - Helv. Chim. Acta, *35*: 1030-1036.

BECK, K. (1912a): Über Kohlensäureausbrüche im Werragebiete der deutschen Kalisalzlagerstätten. - Kali, *6*: 125-128.

BECK, K. (1912b): Petrographisch-geologische Untersuchungen des Salzgebirges im Werra-Fulda-Gebiet der deutschen Kalisalzlagerstätten. - Z. prak. Geol., *20*: 133-158.

BESSERT, F. (1933): Geologisch-petrographische Untersuchungen der Kalilager des Werragebietes. - Archiv für Lagerstättenforschung, H. 57, 45 S., Berlin.

BEST, M.G. (1974): Mantle-derived amphibole within inclusions in alkalic-basaltic lavas. - J. Geophys. Res., *79*: 2107-2113.

BIRCH, F.; CLARK, H. (1940a): The thermal conductivity of rocks and its dependence upon temperature and composition. - Am. J. Sci., *238*: 529-558.

BIRCH, F.; CLARK, H. (1940b): The thermal conductivity of rocks and its dependence upon temperature and composition, Part II. - Am. J. Sci., *238*: 623-635.

BORSTEL, L.E. v. (1989): Stoffbestand einiger Basalte in der Zechsteinfolge 1 (Werra-Fulda-Lagerstättenbezirk). - Diplomarbeit, TU Clausthal.

BOYNTON, W.V. (1984): Cosmochemistry of the rare earth elements: Meteorite studies. - In: HENDERSON, P.: Rare earth element geochemistry, p 63-114, Amsterdam etc. (Elsevier).

BRAITSCH, O. (1962): Entstehung und Stoffbestand der Salzlagerstätten. - 232 S., Berlin-Göttingen-Heidelberg (Springer).

BRAITSCH, O. (1971): Salt deposits, their origin and composition. - Translated by P.J.BUREK and A.E.M.NAIRN in consultation with A.G.HERRMANN and R.EVANS, 297pp, Berlin-Heidelberg-New York (Springer).

BRAITSCH, O., GUNZERT, G., WIMMENAUER, W., THIEL, R. (1964): Über ein Datolithvorkommen am Basaltkontakt im Kaliwerk Buggingen (Südbaden). - Beitr. Mineral. Petrogr., *10*: 111-124.

BREY, G., GREEN, D.H. (1975): The role of CO_2 in the genesis of olivine melilitite. - Contr. Mineral. Petrol., *49*: 93-103.

BROOKINS, D.G. (1981): Geochemical study of a lamprophyre dike near the WIPP site. - In: MOORE, J.G. (ed): Scientific basis for nuclear waste management, *3*: 307-313, New York (Plenum Press).

BROOKINS, D.G. (1984): Geochemical aspects of radioactive waste disposal. - 347pp, New York-Berlin-Heidelberg (Springer).

BROOKINS, D.G. (1986): Natural analogues for radwaste disposal: elemental migration in igneous contact zones. - Chem. Geol., *55*: 337-344.

BRUMSACK, H.-J. (1981): A simple method for the determination of sulfide- and sulfate-sulfur in geological materials by using different temperatures of decomposition. - Fresenius Z. Anal. Chem. *307*: 206-207.

BÜCKING, H. (1881): Basaltische Gesteine aus der Gegend südwestlich vom Thüringer Wald und aus der Rhön. - Jb. kgl. preuß. geol. L.-A. u. Bergakad. f. 1880, S. 149-189.

BÜCKING, H. (1921): Geol. Kte. Preußen u. benachb. dt. Ländern, Blatt Friedewald. - Preuß. Geol. Landesanst., Berlin.

BUNTEBARTH, G. (1980): Geothermie. - 156 S., Berlin-Heidelberg-New York (Springer).

BURNHAM, C.W. (1979): Petrogenesis and the physics of the earth. - In: YODER, H.S. (ed): The evolution of the igneous rocks, p 439-482, Princeton (Princeton University Press).

BYERS, C.D.; GARCIA, M.O.; MUENOW, D.W. (1985): Volatiles in pillow rim glasses from Loihi and Kilauea volcanoes, Hawai. - Geochim. Cosmochim. Acta, *49*: 1887-1896.

CARMICHAEL, I.S.E.; TURNER, F.J.; VERHOOGEN, J. (1974): Igneous Petrology. - 739pp, New York (McGraw Hill).

CARMICHAEL, J.S.E.; NICHOLLS, J.; SPERA, F.J.; WOOD, B.J.; NELSON S.A. (1977): High temperature properties of silicate liquids: applications to the equilibration and ascent of basaltic magma. - Phil. Trans. Roy. Soc. London, *A286*: 373-431.

CARSLAW, H.S.; JAEGER, J.C. (1959): Conduction of heat in solids. - Reprint of 2. ed. (1986), 510pp, Oxford (Clarendon Press).

CAWTHORN, R.G. (1976): Some chemical controls on igneous amphibole compositions. - Geochim. Cosmochim. Acta, *40*: 1319-1328.

CERMAK, V.; RYBACH, L. (1982): Wärmeleitfähigkeit und Wärmekapazität der Minerale und Gesteine. - In: ANGENHEISTER, G. (Hrsg.): Physikalische Eigenschaften der Gesteine, Landolt-Börnstein, Neue Serie, Gruppe V, *1a*: 305-343, Berlin-Heidelberg-New York (Springer).

COX, K.G.; BELL, J.D.; PANKHURST, R.J. (1979): The interpretation of igneous rocks. - London (G. Allen & Unwin).

CREUTZBURG, H. (1965): Bestimmung thermischer Stoffwerte von Salzgesteinen und Nebenge-steinen. - Kali u. Steinsalz, *4*, H. 5: 170-172.

DE LA ROCHE, H.; LETERRIER, J.; GRANDCLAUDE, P.; MARCHAL, M. (1980): A classification of volcanic and plutonic rocks using $R_1 R_2$-diagram and major-element analyses - its relationships with current nomenclature. - Chem. Geol., *29*: 183-210.

DELISLE, G. (1980): Berechnungen zur raumzeitlichen Entwicklung des Temperaturfeldes um ein Endlager für mittel- und hochaktive Abfälle in einer Salzformation. - Z. dt. Geol. Ges., *131*: 461-482.

DIETZ, C. (1928a): Überblick über die Salzlagerstätte des Werra-Kalireviers und Beschreibung der Schächte "Sachsen-Weimar" und "Hattorf" - Z. dt. Geol. Ges., Mb., *1/2*: 68-93.

DIETZ, C. (1928b): Die Salzlagerstätte des Werra-Kaligebietes. - Archiv für Lagerstättenforschung, H. 40, 129 S., Berlin.

DOERFFEL, K. (1967): Die statistische Auswertung von Analysenverfahren und -ergebnissen. - 98 S., Berlin-Heidelberg-New York (Springer).

EDWARDS, A.L. (1966): Trump: A computer program for transient and steady state temperature distributions in multidimensional systems. - Lawrence Radiation Laboratory, Univ. California, Livermore.

EGGLER, D.H. (1974): Volatiles in ultrabasic and derivative rock systems. - Carnegie Inst. Washington Yearb., *73*: 215-224.

FAURE, G.; HOEFS, J.; MENSING, T.M. (1984): Effect of oxygen fugacity on sulfur isotope compositions and magnetite concentrations in the Kirkpatrick Basalt, Mount Falla, Queen Alexandra Range, Antarctica. - Isotope Geosci., *2*: 301-311.

FEELY, H.W.; KULP, J.L. (1957): Origin of gulf coast salt-dome sulphur deposits. - Bull. Am. Assoc. Petrol. Geol., *41*: 1802-1853.

FIELD, C.W. (1972): Sulfur: Element and geochemistry. - In: FAIRBRIDGE, R.W. (ed): The Encyclopedia of Geochemistry and Environmental Sciences, p 1142-1147, New York etc. (Van Nostrand Reinhold Co).

FOURIER, J.B.J (1822): Théorie analytique de la chaleur. - German edition: WEINSTEIN, B.: Analytische Theorie der Wärme, Berlin (1884).

FRANKE, H. (1974): Chemische Untersuchungen an den unter Tage im Grubengebiet "Ernst Thälmann II/III" des VEB Kalikombinat Werra aufgeschlossenen gangförmigen Basalten. - Chemie der Erde, *33*: 188-194.

FRANTZEN, W. (1894): Bericht über neue Erfahrungen beim Kalibergbau in der Umgebung des Thüringer Waldes. - Jb. kgl. preuß. geol. L.-A. u. Bergakad., *15*: 60-61.

FREY, F.A.; GREEN, D.H.; ROY, S.D. (1978): Integrated models of basalt petrogenesis; a study of quartz tholeiites to olivine melilitites from southeastern Australia utilizing geochemical and experimental petrological data. - J. Petrol. *19*: 463-513.

FREERK, M. (1990): Dissertation, Univ. Göttingen, in preparation.

FUHRMANN, U.; LIPPOLT, H.J. (1982): Das Alter des jungen Vulkanismus der Westeifel aufgrund von $^{40}Ar/^{39}Ar$-Datierungen. - Fortschr. Mineral., *60*, Bh. 1: 80-82.

FUKUYAMA, H. (1985): Heat of fusion of basaltic magma. - Earth Planet. Sci. Lett., *73*: 407-414.

GIMM, W. (Hrsg.) (1968): Kali- und Steinsalzbergbau, *1*: Aufschluß und Abbau von Kali- und Steinsalzlagerstätten. - 600 S., Leipzig (VEB Dt. Verlag für Grundstoffindustrie).

GORANSON, R.W. (1942): Sec. 16: Heat capicity; heat of fusion. - In: BIRCH, F., SCHAIRER, J.F., SPICER, H.C.: Handbook of physical constants, repr. 1950, Geol. Soc. Am. Spec. Pap. *36*: 223-242.

GRAMSE, M. (1971): Der Chemismus basaltischer Gläser: Untersuchungen mit der Elektronenmikrosonde. - Fortschr. Mineral., *49*, Bh. 1: 97-98.

GREEN, D.H. (1972): Conditions of melting of basanite magma from garnet peridotite. - Earth Planet. Sci. Lett., *17*: 456-465.

GREEN, D.H. (1973): Experimental melting studies on a model upper mantle composition at high pressure under water-saturated and water-undersaturated conditions. - Earth Planet. Sci. Lett., *19*: 37-53.

GRINENKO, V.A.; DIMITRIEV, L.V.; MIGDISOV, A.A.; SHARAS'KIN, A.Y. (1975): Sulfur contents and isotope composition for igneous and metamorphic rocks from mid-ocean ridges. - Geochem. Int. *12*: 1-132.

GROPP (1919): Gasvorkommen in Kalisalzbergwerken in den Jahren 1907-1917. - Kali, *13*: 33-42, 70-76.

GRUPE, O. (1913): Studien über Scholleneinbrüche und Vulkanausbrüche in der Rhön. - Jb. preuß. geol. Landesanst., *34*: 407-476.

GUTSCHE, A. (1987): Mineralreaktionen, Stofftransporte und Stoffbilanzen im Kontaktbereich Basalt-Sylvinit am Beispiel 1. Begleitflöz im Hangenden des Kalisalzflözes Hessen (K1H), Kaliwerk Hattorf. - Diplomarbeit, Univ. Göttingen.

GUTSCHE, A.; HERRMANN, A.G. (1988): Wechselwirkungen zwischen fluiden Phasen und Evaporiten im Nahbereich von Basaltgängen. - Fortschr. Mineral., *66*, Bh. 1: 49.

HAENEL, R. (1979): Determination of surface temperature in the Federal Republic of Germany on the basis of heat flow values. - Geol. Jb., *E15*: 41-49.

HAJASH, A. (1984): Rare earth element abundances and distribution patterns in hydrothermally altered basalts: experimental results. - Contr. Mineral. Petrol., *85*: 409-412.

HARMON, R.S.; HOEFS, J.; WEDEPOHL, K.H. (1987): Stable isotope (O, H, S) relationships in tertiary basalts and their mantle xenoliths from the northern Hessian Depression, W.-Germany. - Contr. Mineral. Petrol., *95*: 350-369.

HART, S.R.; ALLEGRE, C.J. (1980): Trace-element constraints on magma genesis. - In: HARGRAVES, R.B. (ed): Physics of magmatic processes. p 121-160, Princéton (Princeton Univ. Press).

HARTMANN, G. (1986): Chemische Zusammensetzung und Mineralbestand von Peridotit-Xenolithen mit unterschiedlicher metasomatischer Überprägung aus Basalten der Hessischen Senke. - Dissertation, Univ. Göttingen.

HARTWIG, G. (1954): Zur Kohlensäureführung der Werra- und Fulda-Kalisalzlager. - Kali u. Steinsalz, *1*, H. 5: 3-26.

HASKIN, L.A. (1984): Petrogenetic modelling - use of rare earth elements. - In: HENDERSON, P.: Rare earth element geochemistry, p 115-152, Amsterdam etc. (Elsevier).

HEIDE, K.; BRÜCKNER, U. (1967): Grundlagen zur Phasenanalyse von Salzgesteinen. - Chem. d. Erde, *26*: 235-255.

HEINRICHS, H.; KÖNIG, N.; SCHULTZ, R. (1985): Atom-Absorptions- und Emissionsspektroskopische Bestimmungsmethoden für Haupt- und Spurenelemente in Probelösungen aus Waldökosystem-Untersuchungen. - Berichte des Forschungszentrums Waldökosysteme/Waldsterben, *8*, 92 S., Göttingen (Selbstverlag des Forschungszentrums Waldökosysteme/Waldsterben).

HEINRICHS, H. ; HERRMANN, A.G. (1988): Praktikum der analytischen Geochemie. - 2. überarbeitete u. erweiterte Aufl.: HERRMANN, A.G. (1975): Praktikum der Gesteinsanalyse, Berlin-Heidelberg-New York (Springer).

HENTSCHEL, J.; KLEINITZ, W. (1976): Aufbau der Salzgesteine des Salzstockes Etzel, abgeleitet aus Kernuntersuchungen und Loginterpretationen. - Kali u. Steinsalz, 7, H. 1: 28-39.

HERRMANN, A.G. (1975): Praktikum der Gesteinsanalyse, chemisch-instrumentelle Methoden zur Bestimmung der Hauptkomponenten. - 204 S., Berlin-Heidelberg-New York (Springer).

HERRMANN, A.G. (1979): Geowissenschaftliche Probleme bei der Endlagerung radioaktiver Substanzen in Salzdiapiren Norddeutschlands. - Geol. Rdsch., *68*: 1076-1106.

HERRMANN, A.G. (1980): Geochemische Prozesse in marinen Salzablagerungen: Bedeutung und Konsequenzen für die Endlagerung radioaktiver Substanzen in Salzdiapiren. - Z. dt. geol. Ges., *131*: 433-459.

HERRMANN, A.G. (1983): Radioaktive Abfälle, Probleme und Verantwortung. - 256 S., Berlin-Heidelberg-New York (Springer).

HERRMANN, A.G. (1987): Geochemische Aspekte der Zechsteinevaporite in Deutschland. - Int. Symp. Zechstein 87, Abstracts, S. 40-41, Bochum (Subkommission PERM/TRIAS der Stratigraphischen Kommission DUGW/IUGS).

HERRMANN, A.G. (1988): Gase in marinen Evaporiten. - PTB informiert, *2/88*, 33 S.

HERRMANN, A.G. (1989): Fraktionierungen im Stoffbestand der Zechsteinevaporite Mittel- und Nordwestdeutschlands. - In: Int. Symp. Zechstein 87, (in preparation).

HITE, R.J.; JAPAKASETR, T. (1979): Potash deposits of the Khorat Plateau, Thailand and Laos. - Econ. Geol., 74: 448-458.

HOEFS, J. (1973): Ein Beitrag zur Isotopengeochemie des Kohlenstoffs in magmatischen Gesteinen. - Contr. Mineral. Petrol. *41*: 277-300.

HOEFS, J. (1987): Stable isotope geochemistry. - 3. ed., 241pp, Berlin-Heidelberg-New York (Springer).

HOFFMANN, H.; EMONS, H. (1969): Zur Lösungskinetik des Carnallits unter besonderer Berücksichtigung des Aussolprozesses von Mineralsalzlagerstätten. - Bergakademie *21*: 486-490, 554-558, 674-678.

HOFMANN, A.W.; JACOBY, W.; KRÖNER, A.; LORENZ, V.; WÄNKE, H. (1987): Akkretion und Differentiation des Planeten Erde und ihre Bedeutung für die geodynamische Evolution von Kruste und Mantel. - In: Dt. Forschungsgemeinschaft: Geowissenschaften, Mitteilung XVI: 81-114, Weinheim (VCH Verlagsges.).

HOFRICHTER, E. (1974): Speicherkavernen in Salzstöcken Nordwestdeutschlands - Geologische Probleme, Bemerkungen zur selektiven Auflösung von Kalisalzen. - Erzmetall *27*: 219-226.

HOLLEMAN, A.F.; WIBERG, E. (1971): Lehrbuch der anorganischen Chemie. - 71.-80. Auflage, 1299 S., Berlin (Walter De Gruyter & Co).

HOPPE, W. (1958): Die Bedeutung der geologischen Vorgänge bei der Metamorphose der Werra-Kalisalzlagerstätte. - Freiberger Forschungshefte, *A123*: 41-60.

HOPPE, W. (1960): Die Kali- und Steinsalzlagerstätten des Zechsteins in der DDR, Teil 1: Das Werra-Gebiet. - Freiberger Forschungshefte, *C97*, 166 S.

HUBBERTEN, H.W.; NIELSEN, H.; PUCHELT, H. (1975): The enrichment of ^{34}S in the solfataras of the Nea Kameni Volcano, Santorini Archipelago, Greece. - Chem. Geol. *16*: 197-205.

HULSTON, J.R.; McCABE, W.J. (1961): Mass-spectrometer measurements of the thermal areas of New Zealand, II: Carbon isotopic ratios. - Geochim. Cosmochim. Acta, *26*: 399-410.

HURRLE, H. (1976): Ocelli- und Mandelbildung der ultrabasischen Basalte im Kalisalzlager Buggingen und im Kristallin des Schwarzwaldes. - Jb. geol. Landesamt Baden-Würtemberg, *18*: 19-37.

INGERSOLL, L.R.; ZOBEL, O.J. (1913): Mathematical theory of heat conduciton. - 171pp, Boston (Ginn & Co).

INGERSOLL, L.R.; ZOBEL, O.J.; INGERSOLL, A.C. (1954): Heat conduction. - Rev. ed., 325pp, Univ. Wis. Press.

JACOBS, J.W.; KOROTEV, R.L.; BLANCHARD, D.P.; HASKIN, L.A. (1977): A well-tested procedure for instrumental neutron activation analysis of silicate rocks and minerals. - J. Radioanal. Chem., *40*: 93-114.

JAEGER, J.C. (1957): The Temperature in the neighborhood of a cooling intrusive sheet. - Am. J. Sci., *255*: 306-318.

JAEGER, J.C. (1959): Temperatures outside a cooling intrusive sheet. - Am. J. Sci., *257*: 44-54.

JAEGER, J.C. (1961): The cooling of irregularly shaped igneous bodies. - Am. J. Sci., *259*: 721-734.

JAEGER, J.C. (1964): Thermal Effects of Intrusions. - Rev. Geoph., *2*: 443-466.

JAEGER, J.C. (1968): Cooling and solidification of igneous rocks. - In: HESS, H.H.; POLDERVAART, A. (eds) : Basalts, the Poldervaart treatise on rocks of basaltic composition, *2*: 503-536, New York-London-Sidney (Wiley & Sons).

JAHNE, H.; VOITEL, R.; HAASE, G. (1983): Tektonische Erscheinungsformen im Salinar des Werra-Kaligebietes auf dem Territorium der DDR. - Z. Geol. Wiss., *11*: 1085-1100.

JANDER, G.; BLASIUS, E. (1979): Lehrbuch der analytischen und präparativen anorganischen Chemie (Mit Ausnahme der quantitativen Analyse). - 11. Aufl., 547 S., Stuttgart (S. Hirzel).

JOBMANN, M. (1985): Bestimmung von Wärme- und Temperaturleitfähigkeit an Gesteinen einer Ringintrusion und Berechnung der Abkühlungsgeschichte von Gangintrusionen. - Diplomarbeit, TU Clausthal.

JOCKWER, N. (1981): Die thermische Kristallwasserfreisetzung des Polyhalits und Kieserits in Abhängigkeit von der absoluten Luftfeuchtigkeit. - Kali u. Steinsalz, *8*, H. 4: 126-128.

KÄDING, K.-CH. (1962): Geologische, magnetische und petrographische Untersuchungen tertiärere Vulkanite und ihrer Begleiterscheinungen im Bereich des hessischen Werra-Kaligebietes. - Dissertation, Freie Univ. Berlin.

KÄDING, K.-CH. (1975): Kap.: Zechstein und Kap.: Lagerstätten Kali- und Steinsalz. - In: Erl. Geol. Kte. Hessen, 1:25000, Blatt 5225 Geisa, 2. Aufl., 30-44, 206-207, Wiesbaden (Hess. Landesamt für Bodenforschung).

KÄDING, K.-CH. (1978): Stratigraphische Gliederung des Zechsteins im Werra-Fulda-Becken. - Geol. Jb. Hessen, *106*: 123-130.

KAISER, R.; GOTTSCHALK, G. (1972): Elementare Tests zur Beurteilung von Meßdaten. - Hochschultaschenbuch 774. Mannheim-Wien-Zürich: (Bibliographisches Institut).

KAPPELMEYER, O. (1959): Temperaturfeldmessungen im Grubenbetrieb. - Kali u. Steinsalz, *2*, H. 2: 317-323.

KARSTEN, O. (1954): Lösungsgeschwindigkeit von Natriumchlorid, Kaliumchlorid und Kieserit in Wasser und wässerigen Lösungen. - Z. anorg. allg. Chem. 276: 247-66.

KERN., H.; FRANKE, J.-H. (1980): Thermische Stabilität von Carnallit unter Lagerstättenbedingungen. - Glückauf-Forschungshefte, 41: 252-255.

KNIPPING, B. (1984): Der mineralogische und chemische Stoffbestand am Kontakt zwischen Basalten und Evaporitgesteinen der Werra-Serie des deutschen Zechsteins. - Diplomarbeit, Univ. Göttingen.

KNIPPING, B. (1986): $^{34}S/^{32}S$ ratios of native sulphur in Zechstein 1 evaporites. - Naturwissenschaften, 73: 614.

KNIPPING, B. (1987): Tertiäre Basalte in Perm-Evaporiten (Z1). - Fortschr. Mineral., 65, Bh. 1: 96.

KNIPPING, B. (1989): Basaltische Gesteine in Zechsteinevaporiten. - In: Int. Symp. Zechstein 87, (in preparation).

KNIPPING, B.; HERRMANN, A.G. (1985): Mineralreaktionen und Stofftransporte an einem Kontakt Basalt-Carnallitit im Kalisalzhorizont Thüringen der Werra-Serie des Zechsteins. - Kali u. Steinsalz, 9, H. 4: 111-124.

KOCH, K. (1978): Zur Entstehung von Tonmineralen im Kontaktbereich Basalt-Salinar. - Z. geol. Wiss., 6: 733-747.

KOCH, K.; VOGEL, J. (1980): Zu den Beziehungen von Tektonik, Sylvinitbildung und Basaltintrusion im Werra-Kaligebiet (DDR). - Freiberger Forschungshefte, C347, 104 S.

KOENEN, A. v. (1886): Geologische Specialkarte von Preussen und den Thüringischen Staaten. - 36, Bl. Friedewald; Berlin.

KOENEN, A. v. (1888): Erläuterungen zur geologischen Specialkarte von Preussen und den thüringischen Staaten. - 36, Bl. Friedewald, 16 S.; Berlin.

KOKORSCH, R.; PSOTTA, M. (1984): Die Rohstoffversorgung der Kalifabriken Wintershall und Hattorf. - Kali u. Steinsalz, 9, H2: 39-51.

KÜHN, R. (1951): Nachexkursion im Kaliwerk Hattorf, Phillipsthal - als Beitrag zur Kenntnis der Petrographie des Werra-Kaligebietes. - Fortschr. Mineral., 29/30: 101-114.

KUSHIRO, J.; YODER, H.S. JR.; MYSEN, B.O. (1976): Viscosities of basalt and andesite melts at high pressures. - J. Geophys. Res., 81: 6351-6356.

LAEMMLEN, M. (1975): Geologische Karte von Hessen, 1:25000, Blatt 5225 Geisa. - 272 S., Wiesbaden (Hess. Landesamt Bodenf.).

LAEMMLEN, M.; MEISL, S. (1975): Kap.: Tertiär. - In: Erl. Geol. Kte. Hessen, 1:25000, Blatt 5225 Geisa, 2. Aufl., 108-124, Wiesbaden (Hess. Landesamt für Bodenforschung).

LANE, A.C. (1899): Geology of Isle Royal. - Geol. Survey. Michigan, 6: 106-121.

LANGE, J.; BRUMSACK, H.-J. (1977): Total sulphur analysis in geological and biological materials by coulometric titration following combustion. - Z. Anal. Chem., 286: 361-366.

LARSEN, E.S. JR. (1945): Time required for the crystallization of the great batholith of southern and lower California. - Am. J. Sci., 243A: 399-416.

LEONHARDT, J.; BERDESINSKI, W. (1949/50): Zur laugenfreien Synthese von Salzmineralen. - Fortschr. Mineral., 28: 35-38.

LEVIN, E.M.; ROBBINS, C.R.; McMURDIE, H.F. (1964): Phase diagrams for ceramists. - Columbus, Ohio (Am. Ceramic Soc.).

LIPPOLT, H.J. (1978): K-Ar-Untersuchungen zum Alter des Rhön-Vulkanismus. - Fortschr. Mineral., 56, Bh. 1: 85.

LIPPOLT, H.J. (1982): K/Ar age determinations and the correlation of Tertiary volcanic activity in Central Europe. - Geol. Jb., D52: 113-135.

LOEHR, C.A. (1979): Mineralogical and geochemical effects of basaltic dike intrusion into evaporite sequences near Carlsbad, New Mexico. - M. S. Thesis, Socorro, New Mexico Inst. Mining Technology.

LOVERING, T.S. (1935): Theory of heat conduction applied to geological problems. - Bull. Geol. Soc. Am., 46: 69-94.

LOVERING, T.S. (1936): Heat conduction in dissimilar rocks and the use of thermal models. - Bull. Geol. Soc. Am., *47*: 87-100.

LOVERING, T.S. (1955): Temperatures in and near intrusions. - Econ. Geol., *50*: 249-281.

MAAS, I. (1962): Beiträge zur Isotopengeologie der Elemente Wasserstoff, Kohlenstoff und Sauerstoff. - Isotopentechnik, H. 4: 111-116.

MAGARITZ, M.; SCHULZE, K.-H. (1980): Carbon isotope anomaly of the permian period. - Contr. Sediment. *9*: 269-277.

MAGARITZ, M.; TURNER, P.; KÄDING, K.-CH. (1981): Carbon isotopic change at the base of the Upper Permian Zechstein sequence. - Geol. J., *16*: 243-254.

MAROWSKY, G. (1969): Schwefel-, Kohlenstoff- und Sauerstoff-Isotopenuntersuchungen am Kupferschiefer als Beitrag zur genetischen Deutung. - Contr. Mineral. Petrol., 22: 290-334.

MENGEL, K.; BORSUK, A.; GURBANOV, A.; WEDEPOHL, K.H.; BAUMANN, A.; HOEFS, J. (1987): Origin of spilitic rocks from the southern slope of the Greater Caucasus. - Lithos, *20*, 115-133.

MENNING, M. (1986): Zur Dauer des Zechsteins aus magnetostratigraphischer Sicht. - Z. geol. Wiss., *14*: 395-404.

MESSER, E. (1978): Die nordhessischen Kaligruben. - Kali u. Steinsalz, 7, H. 7: 306-318.

MOORE, J.G.; SCHILLING, J.G. (1973): Vesicles, water and sulfur in Reykjanes Ridge basalts. - Contr. Mineral. Petrol., *41*: 105-118.

MOTTL, M.J.; HOLLAND, H.D. (1978): Chemical exchange during hydrothermal alteration of basalt by sea water - I. Experimental results for major and minor components of sea water. - Ceochim. Cosmochim. Acta, *42*: 1103-1115.

MOTTL, M.J.; HOLLAND, H.D.; CORR, R.F. (1979): Chemical exchange during hydrothermal alteration of basalt by sea water - II. Experimental results for Fe, Mn and sulfur species. - Geochim. Cosmochim. Acta, *43*: 869-884.

MUECKE, G.K. (ed) (1980): Short course in neutron activation analysis in the geosciences. - Short course handbook *5*, 279pp, Mineral. Ass. Can.

MUENOW, D.W.; GRAHAM, D.G.; LIU, N.W.K.; DELANEY, J.R. (1979): The abundance of volatiles in Hawaiian tholeiitic submarine basalts. - Earth Planet. Sci. Lett. *42*: 71-76.

MÜLLER, W. (1958): Über das Auftreten von Kohlensäure im Werra-Kaligebiet. - Freiberger Forschungshefte, *A101*, 99 S.

MUNDRY, E. (1968): Über die Abkühlung magmatischer Körper. - Geol. Jb. *85*: 755-766.

NAUMANN, E. (1911): Über Basaltvorkommen im Salzlager des Schachtes der Gewerkschaft Heldburg. - Z. dt. Geol. Ges., Monatsberichte (März 1910), *62*: 343-344.

NAUMANN, E. (1914): Über einige vulkanische Erscheinungen im Werratale. - Jb. kgl. preuß. geol. L.-A. (1912), *33*, T. 1: 449-467.

NIELSEN, H. (1965): Schwefelisotope im marinen Kreislauf und das $\delta^{34}S$ der früheren Meere. - Geol. Rdsch., *55*: 160-172.

NIELSEN, H. (1981): Schwefelisotope und ihre Aussage zur Entstehung der Schwefelquellen. - In: Die Thermal- und Schwefelwasservorkommen von Bad Gögging, Bayer. Landesamt Wasserwirtsch., H. 15: 99-107.

NIELSEN, H. (1985): Sulfur isotope ratios in strata-bound mineralizations in central Europe. - Geol. Jb., *D70*: 225-262.

NIELSEN, H.; RICKE, W. (1964): Schwefel-Isotopenverhältnisse von Evaporiten aus Deutschland; Ein Beitrag zur Kenntnis von $\delta^{34}S$ im Meerwasser-Sulfat. - Geochim. Cosmochim. Acta, *28*: 577-591.

NOCKOLDS, S.R.; ALLEN, R. (1956): The geochemistry of igneous rock series. - Geochim. Cosmochim. Acta, *9*: 34-77.

NORDLIE, B.E. (1971): The composition of the magmatic gas of Kilauea and its behavior in the near surface environment. - Am. J. Sci., *271*: 417-463.

NORDLIE, B.E. (1972): Gases - volcanic. - In: FAIRBRIDGE, R.W. (ed): The Encyclopedia of Geochemistry and Environmental Sciences, p 387-391, New York etc. (Van Nostrand Reinhold Co).

OEHM, J. (1980): Untersuchungen zu Equilibrierungsbedingungen von Spinell-Peridotit-Einschlüssen aus Basalten der Hessischen Senke. - Dissertation Göttingen.

OEHM, J.; SCHNEIDER, A.; WEDEPOHL, K.H. (1983): Upper mantle rocks from basalts of the northern Hessian Depression (NW Germany). - TMPM Tschermaks Mineral. Petrogr. Mitt., *32*: 25-48.

OELSNER, O. (1961): Ergebnisse neuer Untersuchungen in CO_2-führenden Salzen des Werrareviers. - Freiberger Forschungshefte, *A183*: 5-19.

OESTERLE, F.P.; Lippolt, H.J. (1975): Isotopische Datierung der Langbeinitbildung in der Kalisalzlagerstätte des Fuldabeckens. - Kali u. Steinsalz, *6*, H. 11: 391-398.

PICHLER, H.; SCHMITT-RIEGRAF, C. (1987): Gesteinsbildende Minerale im Dünnschliff. - 230 S., Stuttgart (Enke).

POREDA, R. (1985): Helium-3 and deuterium in back-arc basalts: Lau basin and the Mariana Trough. - Earth Planet. Sci. Lett. *73*: 244-254.

PUCHELT, H.; NIELSEN, H. (1967): Untersuchungen über die Verteilung der Schwefelisotope in den Grubenwässern des Ruhrreviers. - Glückauf-Forschungshefte, H. 6: 303-310.

RAWSON, D.E. (1963): Review and summary of some project Gnome results. - Transact. Am. Geophys. Union, *44*: 129-135.

RAWSON, D.E.; RANDOLPH, R.; BOARDMAN, C.; WHEELER, V. (1966): Post-explosion environment resulting from the Salmon event. - J. Geophys. Res., *71*: 3507-3521.

RAWSON, D.E.; TAYLOR, R.W.; SPRINGER, D.L. (1967): Review of the Salmon experiment, a nuclear explosion in salt. - Naturwissenschaften, H. 20: 525-531.

RINGWOOD, A.E. (1975): Composition and petrology of the earth's mantle. - 618pp, New York (McGraw Hill).

RÖHR, H.U. (1980): Rates of dissolution of salt minerals during leaching caverns in salt - fundamentals and practical application. - In: COOGAN, A.H. & HAUBER, L. (eds): Fifth symposium on salt, p 125-136, Cleveland (The Northern Ohio Geological Society).

ROTH, H.; MESSER, E. (1981): Die Nutzung lagerstättenkundlicher Erkenntnisse für Planung und Betrieb der nordhessischen Kaliwerke. - Kali u. Steinsalz, *8*, H. 5: 145-157.

SAKAI, H.; CASADEVALL, T.J.; MOORE, J.G. (1982): Chemistry and isotope ratios of sulfur in basalts and volcanic gases at Kilauea volcano, Hawaii. - Geochim. Cosmochim. Acta, *46*: 729-738.

SAKAI, H.; DES MARAIS, D.J.; UEDA, A.; MOORE, J.G. (1984): Concentrations and isotope ratios of carbon, nitrogen and sulfur in ocean-floor basalts. - Geochim. Cosmochim. Acta, *48*: 2433-2441.

SAKAI, H.; NAGASAWA, H. (1958): Fractionation of sulphur isotopes in volcanic gases. - Geochim. Cosmochim. Acta, *15*: 32-39.

SCARFE, C.M. (1973): Water solubility in basic and ultrabasic magmas. - Nature, *246*: 9-10.

SCARFE, C.M.; TAKAHASHI, E.; YODER, H.S. JR. (1980): Rates of dissolution of upper mantle minerals in an alkali-olivine basalt melt at high pressures. - Carnegie Inst. Washington Yearb., *79*: 290-296.

SCHEERER, (1911): Gasvorkommen in Kalisalzbergwerken. - Z. Berg- Hütten u. Salinenw., *59*: 212-229.

SCHMIDT, H. (1971): Numerische Lanzeitberechnung instationärer Temperaturfelder mit diskreter Quellenverteilung unter Berücksichtigung temperatur- und ortsabhängiger Stoffwerte. - Dissertation, TH Aachen.

SCHNEIDER, A. (1970): The sulfur isotope composition of basaltic rocks. - Contr. Mineral. Petrol., *25*: 95-124.

SCHNEIDER, A.; NIELSEN, H. (1965): Zur Genese des elementaren Schwefels im Gips von Weenzen (Hils). - Beitr. Mineral. Petrogr. *11*: 705-718.

SCHNEIDER, A.; SCHULZ-DOBRICK, B. (1976): Automatisierte RFA geologischer Proben mit einem Vielkanal-Simultangerät (ARL 72000). - Fortschr. Mineral., *54*, Beih. 1: 150-151.

SDANOWSKI, M.I. (1958): Gesetzmäßigkeiten in der Kinetk der Salzauflösung. - Freiberger Forschungshefte, *A123*: 257-268.

SEYFRIED, JR., W.E.; MOTTL, J. (1982): Hydrothermal alteration of basalt by sea water under sea water-dominated conditions. - Geochim. Cosmochim. Acta, *46*: 985-1002.

SHAW, H.R. (1980): The fracture mechanism of magma transport from the mantle to the surface. - In: HARGRAVES, R.B. (ed): Physics of magmatic processes, p 201-264, Princeton (Princeton Univ. Press).

SHIEH, Y.N.; TAYLOR, H.P. (1969): Oxygen and carbon isotope studies of contact metamorphism of carbonate rocks. - J. Petrol. *10*: 307-331.

SIEMENS, S. (1971): Magnetische ΔT-Messungen im Werra-Kaligebiet zur Erkundung von Basaltgängen. - Kali u. Steinsalz, *5*, H. 11: 385-390.

SMITH, G.D. (1965): Numerical solution of partial differential equations. - London (Oxford University Press).

SMITH, G.D. (1970): Numerische Lösung von partiellen Differentalgleichungen. - 246 S., Braunschweig (Vieweg).

SPERA, F.J. (1980): Aspects of magma transport. - In: HARGRAVES, R.B. (ed): Physics of magmatic processes, p 265-324, Princeton (Princeton Univ. Press).

TAMMANN, G.; Seidel, K. (1932): Zur Kenntnis der Kohlensäureausbrüche in Bergwerken. - Z. anorg. Chem., *205*: 209-229.

TRÖGER, W.E. (1967): Optische Bestimmung der gesteinbildenden Minerale. - 822 S., Stuttgart (E. Schweizerbart).

UEDA, A.; SAKAI, H.; SASAKI, A. (1979): Isotopic composition of volcanic native sulfur from Japan. - Geochem. J., *13*: 269-275.

USDOWSKI, E.; HOEFS, J. (1986): $^{13}C/^{12}C$ partitioning and kinetics of CO_2 absorption by hydroxide buffer solutions. - Earth Planet. Sci. Lett., *80*: 130-134.

VINX, R.; JUNG, D. (1977): Pargasitic-kaersutitic amphibole from a basanitic diatreme at the Rosenberg, north of Kassel (North Germany). - Contr. Mineral. Petrol., *65*: 135-142.

WEBER, K. (1961): Untersuchungen über die Faziesdifferenzierungen, Bildungs- und Umbildungserscheinungen in den beiden Kalilagern des Werra-Fulda Gebietes unter besonderer Berücksichtigung der Vertaubungen. - Dissertation, Bergakademie Clausthal.

WEDEPOHL, K.H. (1963): Die Nickel- und Chromgehalte von basaltischen Gesteinen und deren Olivin-führenden Einschlüssen. - N. Jb. Mineral. Mh., *9/10*: 237-242.

WEDEPOHL, K.H. (1975): The contribution of chemical data to assumptions about the origin of magmas from the mantle. - Fortschr. Mineral., *52*: 141-172.

WEDEPOHL, K.H. (1981): Die primäre Erdmantel- (Mp) und die durch Krustenbildung verarmte Mantelzusammensetzung (Md). - Fortschr. Mineral., *59*, Bh. 1: 203-205.

WEDEPOHL, K.H. (1982): K-Ar-Altersbestimmungen an basaltischen Vulkaniten der nördlichen Hessischen Senke und ihr Beitrag zur Diskussion der Magmengenese. - N. Jb. Mineral. Abh., *144*: 172-196.

WEDEPOHL, K.H. (1983): Die chemische Zusammensetzung der basaltischen Gesteine der nördlichen Hessischen Senke und ihrer Umgebung. - Geol. Jb. Hessen, *111*: 261-302.

WEDEPOHL, K.H. (1985): Origin of the Tertiary basaltic volcanism in the northern Hessian Depression. - Contr. Mineral. Petrol., *89*: 122-143.

WEDEPOHL, K.H. (1987): Kontinentaler Intraplatten-Vulkanismus am Beispiel der tertiären Basalte der Hessischen Senke. - Fortschr. Mineral. *65*: 19-47.

WELZ, B. (1976): Atomic absorption spectroscopy. - 267pp, Weinheim-New York (Verlag Chemie).

WIMMENAUER, W. (1952): Petrographische Untersuchungen über das Ankaratrit-Vorkommen im Kalisalzlager von Buggingen in Baden für 1951. - Mitt.-Bl. bad. geol. Landesanstalt f. 1951, S. 117-129.

WIMMENAUER, W. (1973): Lamprophyre, Semilamprophyre und anchibasaltische Ganggesteine. - Fortschr. Mineral. *51*, H. 1: 3-67.

131

WIMMENAUER, W. (1985): Petrographie der magmatischen und metamorphen Gesteine. - 382 S., Stuttgart (Enke).

WINKLER, H.G.F. (1949a): Kristallgröße und Abkühlung. - Heidelb. Beitr. Mineral. Petrogr., *1*: 86-104.

WINKLER, H.G.F. (1949b): Zusammenhang zwischen Kristallgröße und Salbandabstand bei magmatischen Gang-Intrusionen. - Heidelb. Beitr. Mineral. Petrogr., *1*: 251-268.

WINKLER, H.G.F. (1949c): Crystallization of basaltic magma as recorded by variation of crystal-size in dikes. - Min. Mag. *28*: 557-574.

WINKLER, H.G.F. (1967): Petrogenesis of metamorphic rocks. - 237pp, 2. ed., Berlin-Heidelberg-New York (Springer).

WOHLENBERG, J. (1979): The subsurface temperature field of the Federal Republic of Germany. - Geol. Jb., *E15*: 3-29.

WOHLENBERG, J. (1982): Dichte der Gesteine. - In: ANGENHEISTER, G. (Hrsg.): Physikalische Eigenschaften der Gesteine, Landolt-Börnstein, Neue Serie, Gruppe V, *1a*: 113-119, Berlin-Heidelberg-New York (Springer).

WYLLIE, P.J. (1971): The Dynamic Earth. - 416pp, New York-London-Sydney-Toronto (Wiley & Sons).

YODER, H.S. JR. (1975): Heat of melting of simple systems related to basalts and eclogites. - Carnegie Inst. Washington Yearb., *74*: 515-525.

YODER, H.S.JR. (1976): Generation of basaltic magma. - 256pp, Nat. Acad. Sci., Washington, D.C.

ZIERENBERG, R.A.; SHANKS, W.C.; BISCHOFF, J.L. (1984): Massive sulfide deposit at 21°N, east Pacific rise: chemical composition, stable isotopes, and phase equilibria. - Bull. Geol. Soc. Am., *95*: 922-929.

12 Subject index